Enhancing Mathematics Understanding through Visualization:

The Role of Dynamical Software

Samer Habre
Lebanese American University, Lebanon

Information Science
REFERENCE
An Imprint of IGI Global

Managing Director: Lindsay Johnston
Editorial Director: Joel Gamon
Production Manager: Jennifer Yoder
Publishing Systems Analyst: Adrienne Freeland
Development Editor: Myla Merkel
Assistant Acquisitions Editor: Kayla Wolfe
Typesetter: Christina Henning
Cover Design: Jason Mull

Published in the United States of America by
Information Science Reference (an imprint of IGI Global)
701 E. Chocolate Avenue
Hershey PA 17033
Tel: 717-533-8845
Fax: 717-533-8661
E-mail: cust@igi-global.com
Web site: http://www.igi-global.com

Library of Congress Cataloging-in-Publication Data

Enhancing mathematics understanding through visualization : the role of dynamical software / Samer Habre, editor.
pages cm
Summary: "This book brings these conflicting viewpoints together by offering visual representations as a method of mathematics instruction, exploring the role of technology in providing access to multiple representations of concepts using software applications "-- Provided by publisher.
Includes bibliographical references and index.
ISBN 978-1-4666-4050-4 (hardcover) -- ISBN 978-1-4666-4051-1 (ebook) -- ISBN 978-1-4666-4052-8 (print & perpetual access) 1. Mathematics--Study and teaching--Audio-visual aids. 2. Mathematics--Study and teaching--Data processing. I. Habre, Samer, 1963- editor of compilation.
QA20.C65E54 2013
510.285'5--dc23

2013001600

British Cataloguing in Publication Data
A Cataloguing in Publication record for this book is available from the British Library.

Editorial Advisory Board

Table of Contents

Section 2
The Practice

Section 3
Outside The Norm

Detailed Table of Contents

Section 1
The Theory

Differential equations are the main way in which we make mathematical models of the real world. This is true in just about all fields, from physics to fluid mechanics, from astronomy to finance: if you want to understand how something evolves, or if you want to understand equilibria, you will need to set up and "solve" differential equations. For instance, our understanding of global warming depends mainly on analyzing the differential equations modeling the weather and seeing how their behavior depends on the concentration of greenhouse gasses in the atmosphere. The future of humanity depends on our getting it right.

As described in the communities of practice literature (Lave & Wenger, 1991; Wenger, 1998), boundary objects are material things that interface two or more communities of practice. Extending this, Hoyles, Noss, Kent, and Bakker (2010) defined technology-enhanced boundary objects as, "software tools that adapt or extend symbolic artefacts identified from existing work practice, that are intended to act as boundary objects, for the purposes of employees' learning and enhancing workplace communication" (p. 17). The authors adapt this idea to the undergraduate mathematics classroom and use the phrase "classroom technology-enhanced boundary object" to refer to a piece of software that acts as a boundary object between the classroom community and the mathematical community. They provide three extended examples of these objects as used in a first semester differential equations classroom to illustrate how students' mathematical activity may advance as they interact with the software. These examples show how the applets operate to provide a way for the classroom community to implicitly encounter the mathematical community through the authentic practices of mathematics (Rasmussen, Zandieh, King, & Teppo, 2005). The first example centers on students beginning experience with a tangent vector field applet. The second example develops as the students learn more about solutions to differential equations and leads to a statement of the uniqueness theorem. In the third example, students use a specially designed applet that creates a numerical approximation and its associated image in 3-space relating to a non-technological visualization task that introduces solutions to systems of differential equations.

The author points out that to fully exploit the heuristic potential of Dynamic Geometry Software (DGS) and to increase the heuristic literacy of students, extant DGS teaching units have to be ameliorated in several ways. Thus the author develops a twofold conceptual framework: heuristic reconstruction and heuristic instrumentation of problems. Its origin is rooted in the literature, its use is demonstrated by various examples, and its value is made plausible by an introductory teaching unit and an advanced case study.

Section 2
The Practice

Understanding mathematical concepts is many-folded. Traditional mathematics mostly emphasizes the algebraic/analytical aspect of a problem with minimal reference to its graphical aspect and/or numerical one. In a modern learning environment, however, multiple representations of concepts are proving to be essential for the teaching of mathematics. The availability of user-friendly dynamical software programs has paved the way for a radical yet smooth way for changing the way mathematical concepts are perceived. This chapter presents some of the author's attempts for employing innovative methods in teaching topics in calculus, in differential and difference equations. The focus is on the use of dynamical programs that boost the visual component of the topics being investigated, hence contributing to a more complete understanding of these topics.

Linear differential equations have been well understood for some time and are an important tool for studying the nonlinear systems that most frequently arise in mathematical models of real world systems. Nonlinear systems do not usually have formula solutions, but with graphics, we can see the behaviors of the solutions and thereby "understand" the differential equations. Dynamic and interactive presentations provide students with a streamlined route to understanding these behaviors, resulting in immense power and efficiency that was not previously available at the undergraduate level.

This chapter is inspired by a session the author gave at the 10th EMAC Conference (Engineering, Mathematics, and Applications) at the University of Technology, Sydney, in December 2011. The audience was university teachers, but the software, Autograph, was designed for use in High Schools. The author was able to show how a simple, pedagogically focused interface could be used to create a highly visual approach to the teaching of two favourite topics: Vectors (in 2D and 3D) and Differential Equations (1st and 2nd Order).

This chapter reports on the use of specialized computer applets ("Mathlets") in two different contexts: on-line instruction in calculus through MIT OpenCourseWare and on-campus laboratory exercises on stability of difference schemes in class. Specific applets are described. Three use-cases with varying levels of sophistication (elementary, intermediate, and advanced) are outlined.

The purpose of this chapter is to describe dynamic interactive applets, which are available to all students, to assist in the understanding of sophomore-level engineering mathematics. The topics illustrated here with screen shots are the partial differential equations for heat conduction in one and two space dimensions, the wave equation and Laplace equation in two space dimensions, systems of three ordinary differential equations, and the display of vector fields in three dimensions. An applet is also shown for iterated function systems, although this is not usually part of the engineering sequence. Additional applets mentioned here and available online include the wave equation in one space dimension and two-by-two matrices. The purpose of these applets, all written by the author and freely available, is to help students experiment visually and dynamically with the mathematical concepts.

Modern teaching trends impose the need of spending less time on the manipulative approach to differential and integral calculus, putting the accent on the conceptual understanding of the subject. This chapter presents the standard approach and method used to teach the derivative of a function and indicates some critical points in the teaching of the derivative, offering, at the same time, suggestions for overcoming them. As a supplement, the author gives e-resources that can make possible the implementation of a stimulating, visual, dynamic, and broadened method for teaching the derivative of a function.

The transition from descriptive to inferential statistics is a known area of difficulty for students taking introductory statistics courses. This chapter shares the experiences from a teaching experiment in a college-level introductory statistics classroom that implemented an informal, data-driven approach to statistical inference using the dynamic statistics software Fathom© as an investigation tool. Findings from the study indicate that the informal inferences on which instruction focused in the first part of the course helped students develop understandings of fundamental aspects of inferential and argumentative reasoning that served as foundations for the formal study of inferential statistics in the latter part of the course. The affordances offered by the tool for delving deeply into the data to make sense of the situation at hand were instrumental in supporting student understanding of both informal and formal inferential statistics.

Foreword

Since the1980s, both the practice and the teaching of mathematics have changed. Vast increases in computational power and the accessibility of data have had a dramatic impact on mathematics, greatly expanding its reach. This book examines a central tool in this expansion—dynamic visualization.

Computation has revolutionized mathematical research. It has brought new tools and opened up new areas of application. Explicit computation can now be the most efficient way to verify a conjecture. Visualizations can illuminate mathematical behavior—such a fractals and chaos—leading to new conjectures. Biology, with its myriads of variables, is finally accessible to quantitative analysis. Engineering now has graphical tools at its disposal, and finance has provided a new arena for mathematical modelers. Computer simulation enables drugs to be designed outside the lab. Climate scientists can provide a range of computer-generated scenarios for the planet's future. There is little doubt that computation has changed mathematical research permanently.

With changes in research come changes in the teaching of mathematics. First, topics change. Students going into biology, or finance, or climate modeling, for example, want to learn the mathematics used in their disciplines. They need to understand the mathematics in context—how components of the field are represented in the model, and how changing parameters changes the variables of interest.

At the same time as the changes in computation came changes in communication. The Internet and mobile phones have made data widely accessible. Sales on the Web have put many retailers out of business; data on the Web has made students impatient with contrived applications. Online courses have provided educational access where there was none, and are set to be a central medium of the future.

Changes in computation and communication are poised to affect mathematical pedagogy as profoundly as they have affected mathematical research. Traditionally, one method dominated—"chalk and talk"—and the language used was largely symbolic. Calculus became powerful because of its ability to encapsulate the solution to a problem efficiently in symbols. Thus, for generations students learned mathematics from symbols. Unfortunately, conceptual understanding did not always flow from the symbols. Some students learned to compute, but did not know what they had computed, or how the answer would change if the inputs changed. Classroom norms emerged that equated understanding with symbolic computation. For many students, the phrase "you see that..." came to mean "you know the next step is."

Traditionally, graphical reasoning was avoided by many students because graphs had to be made by hand. Instructors noticed this reluctance when grading student work—unless points were specifically assigned for a graph, many students did not draw one. Many students considered graphing by hand a burden that was not worth the benefits it delivered to understanding.

The arrival of software and graphing calculators in the 1990s tipped the balance. The accessibility of graphs encouraged more students and teachers to use them. At the same time, statistical software was developed to do the number crunching required by real data. This and the ability to make statistical plots enabled exploratory data analysis to be taught in a way that is closer to professional practice. Calculus and differential equations have benefited from powerful new Computer Algebra Systems (CAS) and from increased visualization using software and applets. Thus, the pedagogy in many college-level mathematics courses has been affected by the power of computation, often by creating visualizations.

Not surprisingly, these changes have also generated concerns. The most frequent are about a reduction in computational skill and a loss of familiarity with abstraction. Many mathematicians believe there has been a widespread decline in computational skill; the data suggests that manipulative skills of students who use technology are about the same as the skills of those who do not.

The link between abstraction and visualization is not well understood; this book starts to fill that gap. As one of the defining characteristics of mathematics and one that gives it much of its power, abstraction is of central importance to mathematicians. Yet little is known about how students develop an understanding of abstraction. Everyday teaching suggests that symbols often do not lead students to understand abstraction. However, since many people think visually, there is good reason to believe that visualization will provide an additional, and perhaps effective, bridge to abstraction. The dynamic nature of the visualization may help students take the step from seeing a static connection to seeing the impact of changing variables.

Much of the development of visualization-driven teaching has been done without a grand plan and without systematic reflection. As new software emerged, new ideas for teaching emerged and were tried. Some survived and became part of accepted practice; others died. Probing the relationship between visualization and abstraction, this book is a welcome and timely contribution to the field.

With chapters describing the use of dynamic software in a range of fields, the book gives readers an overview of the possibilities for its use. Most important, this book invites readers to take the next step and investigate the link between visualization and abstraction.

Deborah Hughes Hallett
Cambridge, MA, USA
June 2013

Deborah Hughes Hallett *is Professor of Mathematics at the University of Arizona and Adjunct Professor of Public Policy at the Harvard Kennedy School. With Andrew M. Gleason at Harvard, she organized the Calculus Consortium based at Harvard, which brought together faculty from a wide variety of schools to work on undergraduate curricular issues. She is regularly consulted on the design of curricula and pedagogy for undergraduate mathematics at the national and international levels and has written several college level mathematics texts. She has co-authored a report for the National Academy of Science's Committee on Advanced Study in American High Schools and is a member of the MAA Committee on Mutual Concerns and chair of the College Board's Committee to review the Math-SAT. In 1998 and 2002, she was co-chair of International Conference on the Teaching of Mathematics in Greece, attended by several hundred faculty from about 50 countries. In 2006, she chaired the third conference in this sequence in Istanbul Turkey. She has established programs for master's students at the Harvard Kennedy School, pre-calculus and quantitative reasoning courses (with Andy Gleason), and courses for economics majors. She was awarded the Louise Hay Prize and elected a fellow of the American Association for the Advancement of Science for contributions to mathematics education. Her work has been recognized by prizes from Harvard and the University of Arizona, and as national winner of the MAA Haimo Award for Distinguished Teaching.*

Preface

INTRODUCTION

Understanding mathematical concepts is multi-faceted. Some elements of mathematics are algebraic or analytic; others are numeric; while others are spatial or graphical. In most cases, the three elements are supposed to complement each other. If the balance between the elements is respected then one can expect a (more or less) complete understanding of a particular mathematical concept. In a traditional mathematics setting, however, emphasis is mostly placed on the analytical aspect of a problem without referring to its graphical and/or numerical one. Perhaps one main reason for disregarding the graphical component of a problem is the difficulty associated with this mode of learning and teaching in a pencil-and-paper environment. With the advancement of technology and of computer graphics in particular, the balance can be restored.

In an MAA Notes edited by Walter Zimmermann and Steve Cunningham (1991), the editors spoke of the "visualization renaissance" that has been driven by technological advances. It is now widely accepted that the main strength of technology is in its capability of providing greater and easier access to multiple representations of concepts. But what is visualization? In simple terms, visualization is the learner's ability to represent a mathematical concept graphically via a diagram (be it static or dynamic) and to use the diagram to achieve a more complete understanding of the concept. Visualizing mathematical concepts, therefore, is not an end by itself but rather a means towards an end, and the recent technological advances have largely contributed to a renewed interest in this mode of learning. For many educators and researchers likewise, it is almost impossible to conceive a successful calculus course for example, or a successful course on differential equations, that does not emphasize the visual elements of the subject. As such, the teaching of mathematics in a contemporary classroom has undergone fundamental changes in favor of multiple representations of concepts, in particular in favor of visualization.

The determining role of technology has been an issue of dispute, and the relationship between mathematics, teaching, and technology is a complex one. Educators and researchers have debated about the roles of computers in education. Important questions were raised and investigated, such as: How can technology be effectively used to promote mathematical understanding? What is the role of the teacher in a learning environment that promotes the use of technology, especially if they do not have control over the curriculum they teach? What are ways that enhance the teaching and learning of the current established curriculum? What makes a software program a good one for educational purposes? In a dynamical environment, the user can easily explore via sliders or by dragging; how good is free exploration of mathematical ideas? In addition, one question that has baffled the mathematical community is the effect of these technologies on the curriculum itself. Much of the mathematics curriculum, whether the school or college curriculum was developed at a time when computers were not available. Curriculum developers should be aware that some content may have become obsolete or its importance has diminished. At the same time, topics that were never tackled before can now be handled and taught (such as non-linear systems of ordinary differential equations). These of course have been ruled out in the past on the grounds that time is a constraint or on the grounds that such topics are difficult to teach. Technology, however, can obviate the obstacles that prevented them to be part of current math curricula.

Despite these questions and the lack of definite answers, the actual nature of mathematics teaching has changed considerably because of powerful computers and the "clever mathematical applications on which their operation relies" (Oldknow, et al., 2010). From 3D plots to contour plots, to direction and vector fields, to fractals, one cannot but admire the positive effects of the many mathematical software programs on visualizing many mathematical concepts and ideas, and even sometimes abstract theorems. Whereas in the early days of the visualization renaissance figures were static, now they are dynamic and interactive. With dynamic mathematical software programs, the benefits of a visual representation are doubled to say the least. Dynamic representations of mathematical objects allow learners to visualize problems and mathematical processes in ways that were not possible before. According to Noss and Hoyles (1996), when the algebraic representation is complemented with the dynamic environment, it "comes alive." Learners can view a process as it develops, rather than trying to analyze it from its fixed initial, partial, and end results. Unlike a traditional mathematics education classroom where the learning environment tends to ignore or suppress intuitive or spontaneous ideas, a dynamic environment constitutes a rich experimental and creative arena where students receive the feedback as a direct consequence of their actions and not as a judgmental statement from their teachers.

OBJECTIVE OF THE BOOK AND ITS CONTENTS

Many publications/books have been written about the role of visualization in the teaching and learning of Mathematics. This volume in particular emphasizes the role of dynamical software in enhancing visualization skills and in contributing to a better understanding of certain mathematical concepts. It aims to provide relevant theoretical frameworks and the latest empirical research findings in the area. One issue I had to deal with concerns the static nature of a book, whereas the contributed chapters are supposed to describe the importance of the dynamic nature of the software used in the classroom. In fact, it was an issue that authors had to resolve while preparing their manuscripts, and in my opinion, all did a fine job. It is hoped that an online version of the book be out soon to make up for any lost ideas in the hard copy. Another issue concerns chapters that discuss software developed by the authors. It would be of course inappropriate to include contributions that promote a certain particular software program; the relevant book chapters, however, have focused on the mathematical contents and the pedagogical issues rather than on the software used.

The book is written for practitioners who want to improve their delivery of mathematical concepts and also for researchers in the field. Thus, the potential audience of this book consists of researchers in the area of mathematics education at the college level primarily and of teachers/professors of mathematics who are interested in the improvement of the delivery of the material and in a multi-faceted understanding of mathematical concepts.

The book is divided into three sections:

Section 1: The Theory

This section of the book explores some theoretical backgrounds on the importance of visualization and dynamical software programs in the learning and teaching of mathematics.

In the first chapter, John Hubbard compares and contrasts the standard and the reformed curricula in an ordinary differential equations course. The reformed or "ambitious" curriculum as Hubbard puts it emphasizes visualization and proposes new approaches for the course. K. Keene and C. Rasmussesn in Chapter 2 use the concept of technology-Enhanced Boundary Objects (TEBO) as defined by Hoyles, Noss, Kent, and Bakker (2010) to allow students to experiment and try out new mathematical concepts in conjunction with a set of tasks in differential equations. The authors discuss how students can then reinvent mathematical ideas and gain new knowledge as they interact with the technology. In "Click, Drag, Think! Posing and Exploring Conjectures with Dynamic Geometry Software" by Thomas Gawlick, the author proposes a twofold conceptual framework to fully exploit the heuristic potential of a DGS.

Section 2: The Practice

The second part of the book is experimental. The authors go inside the classroom and describe how a dynamical software program can be exploited to enhance learning.

S. Habre in Chapter 4 presents some of the author's attempts for employing innovative methods in teaching topics in calculus, in differential and difference equations. The focus is on the use of dynamical programs that boost the visual component of the topics being investigated. B. West in Chapter 5 describes how visualization is essential to understand linearization of non-linear systems of differential equations. D. Butler shows in chapter 6 how a simple, pedagogically-focused interface could be used to create a highly visual approach to the teaching of vectors (in 2D and 3D) and differential equations (1st and 2nd Order). In chapter 7, H. Burgiel, C. Lieberman, H. Miller, and K. Willcox of MIT present some interactive applets that they have developed for Calculus and engineering courses. The authors describe the pedagogical factors guiding the creation of this family of applets; the design and usage by students are then presented. More applets for mathematical learning are presented in chapter 8; these applets are designed by Robert Terrell. The derivative concept and how a dynamical software program can be used to explore this concept is presented by Ljubica Dikovic in the next chapter. The last chapter of the second section of the book is concerned with statistical inference. The author, M. Melitiou-Mavrotheris, shares her experiences from a teaching experiment in a college-level introductory statistics classroom which implemented an informal, data-driven approach to statistical inference using Fathom as an investigative tool.

Section 3: Outside the Norm

Rarely do we find similar books that deal with dynamical mathematical software programs and cross-curricular issues such as art and literature.

In Chapter 11, B. Felsager focuses on the use of such software programs when teaching the concept of infinity in mathematics as well as in literature. In Chapter 12, S. Habre makes a second contribution where linear iterative systems and the dynamical software program discussed in Chapter 4 are used to create a popular and old art called string art.

Samer Habre
Lebanese American University, Lebanon

REFERENCES

Noss, R., & Hoyles, C. (1996). *Windows on mathematical meanings: Learning cultures and computers*. Dordrecht, The Netherlands: Kluwer. doi:10.1007/978-94-009-1696-8.

Oldknow, A., Taylor, R., & Telow, L. (2010). *Teaching mathematics using ICT* (3rd ed.). London: Continuum International Publishing Group.

Zimmermann, W., & Cunningham, S. (1991). *Visualization in teaching and learning mathematics*. Washington, DC: Mathematical Association of America.

Acknowledgment

In the end, I wish to thank all the contributors and the chapter reviewers. I hope that readers will find this volume useful and find ideas that are useful for their research work and in the classrooms. I hope readers will be encouraged to use visualization means in their classes if they have not done so already. The opportunities for creativity in a dynamical environment seem to be boundless.

Section 1
The Theory

Chapter 1
Technology and Differential Equations

John Hubbard
Cornell University, USA

ABSTRACT

Differential equations are the main way in which we make mathematical models of the real world. This is true in just about all fields, from physics to fluid mechanics, from astronomy to finance: if you want to understand how something evolves, or if you want to understand equilibria, you will need to set up and "solve" differential equations. For instance, our understanding of global warming depends mainly on analyzing the differential equations modeling the weather and seeing how their behavior depends on the concentration of greenhouse gasses in the atmosphere. The future of humanity depends on our getting it right.

THE STANDARD CURRICULUM

This grandiose view of the central importance of differential equations in how we understand the world is not much in evidence in courses on differential equations, particularly not in elementary courses.

The First Course

The students are likely to encounter a bag of tricks to find solutions to various equations, especially first order separable equations and second-order linear differential equations with constant coefficients. These are of course important topics.

DOI: 10.4018/978-1-4666-4050-4.ch001

An explicit solution is a wonderful thing, especially if the formula is analyzed and graphed to emphasize the behavior of the solution. For linear differential equations with constant coefficients, understanding how the signs of the real parts of eigenvalues affect the behavior of solutions is critical in many situations.

I fear that often far less attention is focused on the meaning of differential equations than on obtaining the formulas.

A More Ambitious Course

In a more ambitious first course, the students may well see in addition a proof of the existence and uniqueness theorem, probably based on Picard iteration, and solutions of the heat equation and the wave equation based on Fourier series, using separation of variables. There might also be an introduction to numerical methods, usually independent of the existence and uniqueness theory.

Again, these are important topics; but I have reservations about how they are taught in practice. The Fourier solutions to the heat and wave equations are really the first chapter of the spectral theory of differential operators, and the fact that separation of variables leads to eigenvalues and eigenfunctions, closely related to linear equations with constant coefficients, is often overlooked.

My reservations about existence and uniqueness are of a different nature. They are based on a very disturbing observation: I often had the experience of asking students questions like "does the solution of $x' = x^2 - t$ with $x(0) = x_0$ exist for all $t \geq 0$?" The students would fiddle with the equation for a while, try various changes of variables, and eventually come up with the conclusion that "this equation has no solutions." These are students who could probably state the existence and uniqueness theorem, and perhaps prove it. But the "solution" provided as a limit of Picard iterations did not have the necessary psychological reality in the student's minds for them to think about it effectively.

A MORE APPROPRIATE APPROACH

With Beverly West I developed during the 1980's and 90's a different curriculum, embodied in the books *Differential Equations, A Dynamical Systems Approach, Parts I and II* (Hubbard & West, 1991, 1995), and the software package *MacMath* (Hubbard & West, 1992, 1999)

Principle 1: Solutions to differential equations are functions, and the object of a course on differential equations is to describe the behavior of such functions.

By behavior, I mean answers to such questions as:

1. Determine or approximate the domain of existence of solutions.
2. How do the solutions behave at the boundary of the domain of existence: Do they have vertical asymptotes, do they oscillate, etc.?
3. Where are they increasing or decreasing? Where do they have maxima and minima?

The solutions are usually not expressible in elementary terms or as indefinite integrals of elementary functions, and the course should provide tools to describe the behavior of solutions without finding explicit solutions.

An important example of this phenomenon occurs already before the first course on differential equations. Consider the differential equation $x' = ax,$ the model of compound interest and the simplest model of population growth and radioactive decay. If your bag of "known equations" only includes rational functions (the only functions whose values can be computed using arithmetic), then $x' = ax$ defines the new function $C\exp(ax),$ and the best way to study these functions is as solutions of the differential equation. Trying to define e^x as the number e (whatever it is) raised to the power *x* (whatever that means: what is $2^{\sqrt{2}}$?) is much more difficult than understanding the differential equation. It is only by convention that elementary means "combinations of rational, exponential, logarithmic and trigonometric functions." Each of these is introduced as a solution of an appropriate differential equation, and one could include in one's repertoire elliptic functions, Bessel functions, etc.

Fences and Funnels

The most innovative aspect of our approach may be the introduction of *fences* and *funnels*, which very often allow the user to analyze scalar differential equations $x' = f\left(t, x\right)$ and their corresponding slope fields.

Our standard example is $x' = x^2 - t,$ in part because it is important in several applications (it is the Riccati equation associated to the Airy equation which is used in optics and hydrodynamics; the behavior of solutions to Airy's equation is analyzed in Noonburg et al, 1994 and 1997), and in part because it provably has no solutions that can be written in elementary terms, or indefinite integrals of elementary functions, or indefinite integrals of those, etc. (Hubbard & Lundell, 2011).

Not bad for no formulas, eh!

But it is quite easy to describe the solutions (See Figure 1), and not too difficult using fences and funnels to prove that the description is correct. See Hubbard and West (1995).

Figure 1. This picture illustrates the differential equation $x' = x^2 - t$, *and how fences, funnels and antifunnels help to analyze the solutions. The region* $\sqrt{t} \le x \le \sqrt{t} + 1$ *(lightly shaded) is an antifunnel, containing a unique solution* u_0 *asymptotic to* $+\sqrt{t}$ *as. The solution* u_0 *(dark curve) is defined for* $A < t < \infty$ *for some number* $A \approx -2.3$; *it is increasing throughout its domain, and has the vertical asymptote* $t = A$ *to the left (also a dark line). All solutions whose graphs lie above the graph of* u_0 *are increasing, have bounded domains of existence, and have vertical asymptotes both to the left and the right. All solutions whose graphs lie below the graph of* u_0 *have a vertical asymptote to the left, increase until they cross the curve* $t = x^2$, *then decrease, enter the funnel* $-\sqrt{t} \le x \le -\sqrt{t} - 1, t \ge \frac{5}{4}$ *(shaded a bit darker), and are asymptotic to* $-\sqrt{t}$ *as* $t \to \infty$.

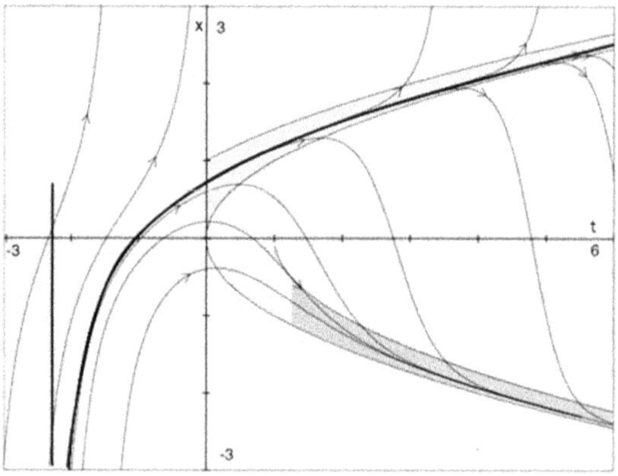

We found the program *diffEq,* one of the programs of *MacMath,* absolutely essential in bringing the students to the point where they could even think of "describing the solutions of a differential equation." High-school math is so number- and formula-oriented that making sentences about anything mathematical does not come easily to students. In any case, using a combination of the software and the fence and funnel techniques, we really overcame the psychological barrier of thinking about differential equations without formulas.

Formulas After All

Almost any function is a fence of some sort in specific ranges, but the really interesting ones come from comparing one differential equation to another, perhaps one that can be solved.

For instance when $|x|$ is large, solutions to $x' = x^2 - t$ should behave much like solutions to $x' = x^2$ and since these explode we should expect the solutions of $x' = x^2 - t$ to explode also, at least some of them. So we discuss the most useful of the several classes of solvable differential equations, especially separable and linear ones, with emphasis on the behavior of solutions.

Numerical Methods

At some point, we need to put some solid ground under our feet: to know that the solutions we are airily describing actually exist. We carry this out in a way compatible with Principle 2.

Principle 2: The best proof of existence and uniqueness, in any setting, is to prove that an effective approximation procedure converges.

In order to stay faithful to this principle, we have found it sufficient to introduce various elementary (but perfectly effective) numerical methods: Euler, mid-point Euler, 4th order Runge- Kutta in particular. Here again software is essential to see how these methods work in practice, what sorts of convergence one finds. The *MacMath* program *Numerical Methods* was essential to convey what the various methods do, how they can behave and misbehave.

Existence and Uniqueness

We prove existence and uniqueness by showing that Euler's method does indeed converge when the step length becomes small, at least when the differential equation is Lipschitz. We also emphasize the meaning of the Lipschitz constant: the rate at which solutions pull apart (forwards or backwards). If there is no Lipschitz constant, this rate is infinite, leading to non-uniqueness. Our Fundamental inequality, really a variant of Gronwall's inequality, embodies these ideas. We prove it by a fence argument. A paper by with Habre, Hubbard, and West (2001) illustrates the wild things that can happen when a differential equation does not satisfy a Lipschitz condition.

Iteration

Every numerical method is an iteration, so it is helpful to learn to understand some iterations. We are concerned with the question: "What happens in the long run for a particular numerical method?" rather than the usual question, "what happens at a fixed time as the step-length goes to 0 (which is largely answered in the previous section)?"

Of particular importance is the notion of stiffness. This is usually encountered when studying partial differential equations called reaction-diffusion equations: the heat equation (diffusion) perturbed by a non-linear term (the reaction). Numerical analysts long ago discovered a most unpleasant fact: as such, an equation is discretized more and more finely, all direct numerical methods become unstable, as illustrated in Figure 2. Equation $x' = ax$ exhibits the same behavior: as a becomes large, so that the origin is more and more stable, numerical methods for a fixed step length become unstable. For reaction-diffusion equations, as you discretize more and more finely, you "see" more and more eigenvalues of the Laplacian, with more and more negative eigenvalues. These behave like the a in $x' = ax$; when they are sufficiently negative, all direct methods become unstable.

Higher Dimensions

Few systems can really be modeled by scalar first order equations. Examples of models in higher dimensions are: Volterra-Lotka in population dynamics, and Newton's derivation of Kepler's laws from the law of gravitation. By adding variables,

Figure 2. We see "approximations" to the solutions of $x' = x^2 - t$, drawn by Euler's method (thick grey line), midpoint Euler (thick black line) and 4th order Runge-Kutta (thin black line). As you go down the funnel $-\sqrt{t} \leq x \leq -\sqrt{t} - 1$, the solutions become more and more stable, but the numerical approximations go berserk. Sooner or later, all three methods (and all direct methods) develop wild oscillations. The behavior of mid-point Euler is particularly worrisome: for quite a while, the approximation looks like a perfectly reasonable curve, but this curve is an artifact of the numerical method and has nothing to do with the differential equation.

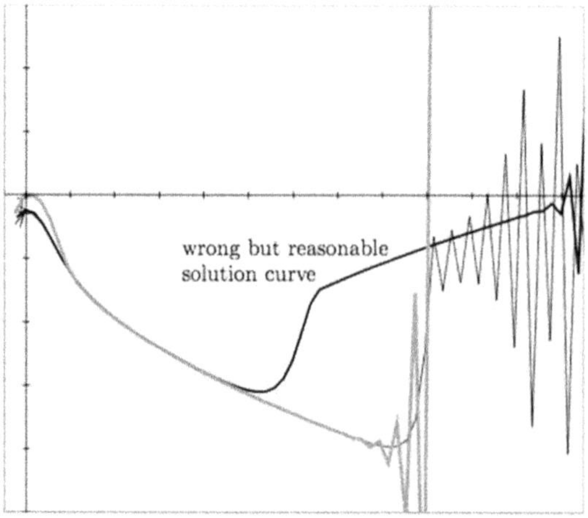

one can transform higher order equations into first order equations, and advocate switching systematically to first order systems rather than higher order scalar differential equations.

Linear Equations and Linearization

We show that the theory of linear differential equations with constant coefficients is essentially the same thing as the theory of eigenvalues and eigenvectors. In particular, we have Principle 3.

Principle 3: The behavior of the solutions to a system of differential equations is controlled by the sign of the real parts of the eigenvalues.

This really comes to life when studying non-linear equations and linearizing them near their zeros. The non-linear analog of the eigenvectors are the *separatrices*, which separate regions of different behavior. To this day understanding separatrices (also called stable and unstable manifolds) is the most powerful tool we have to study non-linear equations.

THE FORCED DAMPED PENDULUM

We next give an application of the ideas in the previous section. One reason is to illustrate the superiority of these simple ideas to those involved in any standard curriculum, even if quite advanced with all sorts of asymptotic developments and expressions of solutions as integrals depending on a parameter.

Consider the forced damped pendulum, described by the differential equation:

$$x'' + ax' + b\sin x = c\cos 2\pi\omega t \tag{1}$$

This is a "garden-variety" equation; people undoubtedly studied it in the 18th and 19th centuries. They had no idea what a monster they were looking at. What I will describe concerns mainly the values $a = 1, b = 1, c = 1, \omega = \dfrac{1}{2\pi}$.

Forced damped pendulums are fundamental throughout engineering: every robot arm is a forced damped pendulum. Analyzing its behavior involves symbolic dynamics and *Smale horseshoes*, which are the next step in the curriculum described in Section 2. More detailed analyses of the mathematics discussed here are given in Hubbard

(1999 and 2000). Such problems and methods have traditionally been totally beyond any undergraduate (or even graduate) curriculum. But with appropriate software they become a challenging but approachable problem at the undergraduate level.

The first step is to change the second-order Equation (1) into a system of first order equations:

$$x' = y \tag{2}$$

$$y' = -ay - b\sin x + c\cos 2\pi\omega t \tag{3}$$

The second step is to graph various solutions (See Figure 3). The pictures are pretty bewildering at first, but after a while one sees what is going on: in these variables, where $x \in R$ represents an angular variable, there are infinitely many sinks whose x-coordinates differ by multiples of 2π. Most solutions appear to eventually settle down to one of these.

This leads to the third, and probably most important step: to color each initial condition, say at time 0, according to which periodic solution the corresponding solution settles down. Thus, we obtain Figure 4.

Figure 3. We see 16 solutions $t \to x(t)$ of the differential equation (1), all with very nearly the same initial condition $x(0) \approx 1$, $x'(0) = 0$. They are plotted for time $0 \le t \le 50$. They appear to go through wild gyrations, before settling down, at least some of them, to periodic solutions. These solutions, denoted . . . , S_{-2}, S_{-1}, S_{-0}, S_1, . . . are really all the same motion -- they differ by multiples of 2π. What makes them different is the number of times the pendulum goes over the top before settling down.

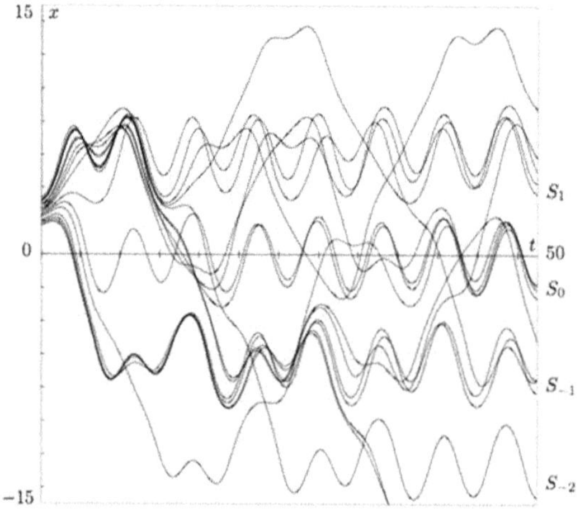

Figure 4. This picture would be better in color. It shows the basins of the motions $\ldots s_{-2}, s_{-1}, s_0, s_1, s_2, \ldots$; *there actually are infinitely many basins, repeating periodically. As the reader can see, the basins are tremendously intermingled; each consists of an "octopus" with four arms, which spread out and wander. They intermingle so much that any point on the boundary of one basin is on the boundary of all (infinitely many) others.*

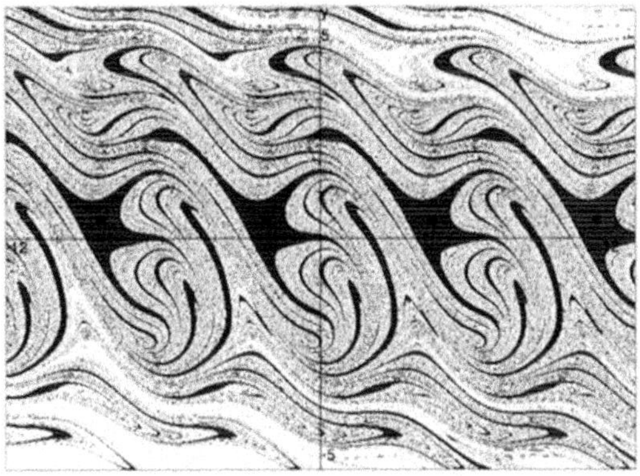

Figure 4 is obviously very rich, and invites analysis.

Finally, let us state two theorems that can be proved about this figure.

Theorem 1: The basins form lakes of Wada: every point in the boundary of one basin is in the boundary of all the others.

Note that this is a way of saying that the pendulum modeled by equation (1) is chaotic; if you perturb an initial condition on thc boundary of a basin by an arbitrary small amount, you can make it go anywhere.

The method of proof is due to Nusse and Yorke (1996). It involves analyzing stable and unstable separatrices of appropriate periodic motions of the pendulum. It would presumably be impossible to prove without computers and appropriate software (in the sense that no human could carry out the amount of computation needed for the proof). The computation is trivial for a modern computer, but even with the computer, the proof is fairly hard.

The second theorem is easier and presumably more important because the sort of chaos it describes is easier to analyze, and has applications in many fields. This result follows from exhibiting a *Smale horseshoe* in the dynamical plane. *Smale horseshoes* go back to the work of Smale (1967). It is so important that there are hundreds if not thousands of works that use the methods.

The differential equation (1) has a natural unit of time: the period of the forcing term $c\cos\omega t$ c cos ωt, i.e. $T = \frac{1}{\omega}$. During one unit of time, the pendulum must do one of four things:

1. Cross the bottom position exactly once counterclockwise (event +1).
2. Cross the bottom position exactly once clockwise (event −1).
3. Not cross the bottom position (event 0).
4. None of the above (event NA) (for instance crossing the bottom position twice).

Define the *k*-th time interval to be $kT \le t \le (k+1)T$.

Theorem 2: Choose any biinfinite sequence.

$$\ldots s_{-2}, s_{-1}, s_0, s_1, s_2, \ldots$$

of the symbols −1,0,1. Then there is a state $(x(0), y(0))$ *of the pendulum at time 0 such that the corresponding motion* $(x(t), y(t))$ *of the pendulum will "do"* s_k *during the k-th interval of time.*

This is also a way of saying that the pendulum is chaotic: by choosing an appropriate initial condition, you can make it do anything you like.

Here the point is that any time you can exhibit a *Smale horseshoe* for a mapping or a differential equation, a similar result will follow. As it turns out, such horseshoes are common, but exhibiting them is usually impossible without the computing power of a computer.

More importantly, systems that contain such horseshoes can be *controlled*: with only small expense, the system can be perturbed so as to set it on "any" course one wishes.

REFERENCES

Habre, S. S., Hubbard, J. H., & West, B. H. (2001). The convergence of an Euler approximation of an initial value problem is not always obvious. *The American Mathematical Monthly*, *108*(4), 326–335. doi:10.2307/2695239.

Hubbard, J. H. (1999). The forced damped pendulum: chaos, complication and control. *The American Mathematical Monthly*, *106*, 741–758. doi:10.2307/2589021.

Hubbard, J. H. (2008). *The forced damped pendulum: Chaos, complication and control*. Retrieved from http://www.math.cornell.edu/~hubbard/Wisconsinpendulumtalk.pdf

Hubbard, J. H., & Lundell, B. E. (2011). A first look at differential algebra. *The American Mathematical Monthly*, *118*, 245–261. doi:10.4169/amer.math.monthly.118.03.245.

Hubbard, J. H., McDill, J. M., Noonburg, A., & West, B. H. (1994). A new look at the airy equation with fences and funnels. *The College Mathematics Journal*, *25*(5), 419–431. doi:10.2307/2687507.

Hubbard, J.H., & West, B.H. (1991). Differential equations: A dynamical systems approach: Part I: Ordinary differential equations. *Texts in Applied Mathematics,* (5).

Hubbard, J. H., & West, B. H. (1992). *MacMath 9.2*. New York: Springer-Verlag.

Hubbard, J.H., & West, B.H. (1995). Differential equations: A dynamical systems approach: Part II: Higher-dimensional systems. *Texts in Applied Mathematics,* (18).

Hubbard, J.H., & West, B.H. (1999). *MacMath 10.0*. Unpublished.

Nusse, H. E., & Yorke, J. (1996). Wada basin boundaries and basin cells. *Physica D. Nonlinear Phenomena*, *90*(3), 242–261. doi:10.1016/0167-2789(95)00249-9.

Smale, S. (1967). Differentiable dynamical systems. *Bulletin of the American Mathematical Society*, *73*(6), 747–817. doi:10.1090/S0002-9904-1967-11798-1.

Chapter 2

Sometimes Less is More:
Examples of Student-Centered Technology as Boundary Objects in Differential Equations

Karen Keene
North Carolina State University, USA

Chris Rasmussen
San Diego State University, USA

ABSTRACT

As described in the communities of practice literature (Lave & Wenger, 1991; Wenger, 1998), boundary objects are material things that interface two or more communities of practice. Extending this, Hoyles, Noss, Kent, and Bakker (2010) defined technology-enhanced boundary objects as, "software tools that adapt or extend symbolic artefacts identified from existing work practice, that are intended to act as boundary objects, for the purposes of employees' learning and enhancing workplace communication" (p. 17). The authors adapt this idea to the undergraduate mathematics classroom and use the phrase "classroom technology-enhanced boundary object" to refer to a piece of software that acts as a boundary object between the classroom community and the mathematical community. They provide three extended examples of these objects as used in a first semester differential equations classroom to illustrate how students' mathematical activity may advance as they interact with the software. These examples show how the applets operate to provide a way for the classroom community to implicitly encounter the mathematical

DOI: 10.4018/978-1-4666-4050-4.ch002

community through the authentic practices of mathematics (Rasmussen, Zandieh, King, & Teppo, 2005). The first example centers on students beginning experience with a tangent vector field applet. The second example develops as the students learn more about solutions to differential equations and leads to a statement of the uniqueness theorem. In the third example, students use a specially designed applet that creates a numerical approximation and its associated image in 3-space relating to a non-technological visualization task that introduces solutions to systems of differential equations.

INTRODUCTION

Visualization offers students a window into the world of mathematics. Technology used in mathematics at the university level plays many roles in facilitating visualization. For example, in beginning calculus many classes use software such as Mathematica, Matlab, or Maple to help students to visualize graphs of functions, derivatives, integrals, and series. Animations are often available for use with these programs, and instructors or programmers may set up demonstrations to show students a visual illustration to help them understand the mathematics. Additionally, the development of Java applets to illustrate a specific idea in calculus, differential equations, non-Euclidean geometry, and other courses can provide an opportunity for students to visualize the intended mathematics. Over the past several years, we have been engaged in designing and studying innovative learning environments in differential equations that provide students with significant opportunities to visualize concepts as they reinvent key mathematical ideas (Rasmussen & Kwon, 2007). The use of specially designed technological tools has been a central component of these efforts. The purpose of this chapter is to illustrate how these technological tools function as boundary objects between the classroom community and the broader mathematical community.

As illustrated in the communities of practice literature (Lave & Wenger, 1991; Wenger, 1998), boundary objects are material things that interface two or more communities of practice and satisfy the informational requirements of all of them (Bowker & Star, 1999; Lee, 2007). A community of practice is a collective construct in which the joint enterprise of achieving particular goals, such as doing mathematics, evolves within the social connections of that particular group (Lave & Wenger, 1991; Wenger, 1998). Boundary objects can be mathematical symbols, technology, documents, software, or other items that allow people to cross between different communities and work together. Hoyles, Noss, Kent, Bakker (2010) define Technology-Enhanced Boundary Objects (TEBOs) as, "software tools that adapt or extend symbolic artifacts identified from existing work practice, that are intended to act as boundary objects, for the purposes of employees' learning and enhancing workplace communication" (p. 17). We adapt this idea, which was actually developed to study learning in the workplace, to the undergraduate mathematics classroom and hence use the phrase "classroom TEBO" to refer to a piece of software that acts as a boundary object between the classroom community and the mathematical community. Classroom TEBOs serve the purpose of promoting student learning and enculturation into the discipline of mathematics. It is important to note that a classroom TEBO is only a boundary object as the students' interact with it, other students, and the teacher, who in this case is a representative of the mathematics community.

As mathematics instructors may have noticed, most off-the-shelf software tools illustrate the mathematics in more or less the form used by experts. For example, when a solution to a differential equation is given, expert software typically allows students to chose an initial condition and then the graph (or a close approximate) is shown. Analytic solutions, when they exist, might also be provided. There is no need for students to consider the form or context of the differential equation as the complete answer is there in front of them. Moreover, students tend to be passive observers. This might be fine for experts, as they have likely solved the problem for themselves and have deep understanding as the basis for the answer. When they see the answer, there is already a firm background that supports the answer. After students use this type of software, however, they may not have anything more than the answer! Typical examples of such traditional use of technology in education include those from Perge (2007), who shows how to use Maple to plot solutions to a second order differential equation that models a forced pendulum, and those of Maat and Zakaria (2010), who use Maple to symbolically solve differential equations.

In contrast to these more traditional uses of software, classroom TEBOs are technological tools that are designed to be less sophisticated than the traditional software and to allow students to experiment and try out new mathematical concepts in conjunction with a set of tasks. Students can then reinvent mathematical ideas as they interact with the technology and experience the mathematics to gain new and/or more sophisticated knowledge. The classroom TEBOs we illustrate do not do "everything" like some software. Instead, they provide a "playground" for students to work in which allows the students to actively reinvent the mathematical ideas (Gravemeijer, 1994). Classroom TEBOs are a "less is more" type of technology because they provide less capability that are required by an expert. As such, they allow more learning to occur as students reinvent important mathematical ideas.

We provide three extended examples of classroom TEBOs to illustrate how students' mathematical activity may advance during as they interact with the software. Additionally, these examples illustrate the concept of boundary object as we see it emerge during the activity. Each of these examples occur in a first semester student-centered differential equations class and involves what one might call minimalist applets in that they are designed for the students to bring their own mathematical notions to the experience. These examples show how the applets operate as classroom TEBOs, where the boundary object provides a way for the classroom community to implicitly encounter the mathematical community through the authentic practices of mathematics (Rasmussen, Zandieh, King, & Teppo, 2005).

The first example centers on students beginning experience with a tangent vector field applet. This applet shows a coordinate axis with tangents plotted after being provided a differential equation. The students can use the cursor to "move" a point and its associated tangent vector around and imagine the approximate behavior of the solution functions for the given differential equation. The applet does not sketch in a solution, or provide any type of analytic solution (things an expert would require). One part of the instructional activity provides a way for the students to sketch solutions using transparencies placed on the computer screen for the students to physically enact the emergence of a solution from a tangent vector field. The students work with this applet to reinvent Euler's method and they encounter the mathematical culture as they develop intuition and understanding about how the flow of the vector field "determines" solutions to a differential equation.

The second example develops as the students learn more about solutions to differential equations. Developing intuitions for why certain conditions for theorems are what they are is something that experts appreciate and seek out. Students therefore, as they become enculturated into the discipline of mathematics, need such experiences. In the classroom TEBO that leads to a statement of the uniqueness theorem, students experiment with a tangent vector field applet to investigate how a solution to dh/dt=-h and dh/dt= -h1/3 creates different possible imagined paths for a helicopter landing. The different behavior of the solutions at h=0 are explored and the students develop intuition for the conditions of a version of the uniqueness theorem in terms of the boundedness of the partial derivative with respect to h. The visual experience with the tangent vector field dynamic behavior is key, as it leads to observations about how the vector "snaps" to the t-axis as h decreases to zero. The teacher, as a representative of the mathematical community, then helps students connect their work and observations to the formal statement of the theorem.

The third example comes at the beginning of the instructional sequence for systems of differential equations. Students first work with an imaginary situation that mimics what the physical trace of a dot at the end of a propeller would look like on a slow moving airplane on the ground moving in a straight line. They consider the view of this trace from several different perspectives. Then students use a specially designed applet that creates a numerical approximation and its associated image of in 3-space, both of which represent an approximate solution. The students explore this curve by rotating it around using the mouse to visualize what the curve looks like from different perspectives, similar to that of the airplane task. Students often call the three dimensional solution curve the "mother curve" that all the other solution views are determined by. Thus, this second applet serves as a classroom TEBO for the students to develop mathematical understanding about solutions to systems and how they appear from different points of view and in 3-space. Students work with this TEBO then motivates the need for the phase plane and vector fields.

Before we describe the three research based classroom TEBO examples, we offer a brief research review in three areas: Research on the community of practice as a classroom construct and a construct for the mathematical community, research published on the notion of boundary objects, and research on using technology as a visualization tool in differential equations.

SUPPORTING AND DEFINING LITERATURE

Communities of Practice

A community of practice refers to a collective learning environment in which, over time, practices that reflect the pursuit of enterprises, such as becoming a mathematician, evolve within the social structures of that particular group (Lave & Wenger, 1991; Wenger, 1998). The construct of a community of practice has been used in mathematics education to describe both classroom communities of learning and the mathematician community (i.e. Boaler, 2002; Burton, 1999; Rasmussen, Zandieh, & Wawro, 2009). This is not the only way to describe these communities, but it does provides a way to discuss the social features of learning and the ways in which students participate in the classroom community and the broader mathematical community.

Burton (1999, 2004) provides a view into the life and work of mathematicians. She describes mathematicians in England as a male community of practice and looks formally at the practice of mathematics and what it means to know mathematics from a mathematician's point of view. Herzig (2004) documents mathematicians as a community of practice as a way to address the issues of women and people of color in mathematics communities. Lambertus (2009) provides a more particular view of mathematicians in a specific mathematics department in a large United States university and documents it as a community of practice. Solomon (2007), on the other hand, argues that undergraduates often do not feel part of the mathematics community of practice that exists at a university. Solomon's work raises the question of what can be done to help students feel more aligned with the practices of mathematicians. The notion of classroom TEBOs developed in this chapter offers a start to addressing this problem.

As far as classrooms are concerned, several researchers have explicated that some classrooms do indeed operate as a community of practice. This research tends to be grounded in mathematics classrooms based on inquiry where teachers serve as facilitators and learning occurs by students discussing their own ideas. Goos (2004) called such classroom communities of practice a "community of inquiry" and she documents one such example in an Australian secondary mathematics classroom. Goos work uses Vygotsky's notion of zone of proximal development (ZPD) as an analytic tool for describing how learning can be explained as increasing participation in a community of practice. While we do not explicitly take up the ZPD as an analytic tool in our analysis, this idea is certainly consistent in how we analyze the role of TEBOs as tools that scaffold students as they progress from their current ways of thinking to those that are more compatible with the discipline of mathematics. Also from a Vygotskian perspective, Lerhman (2001) and van Oers (2001) both discuss the nature of classrooms that are communities of learners where students are enculturated into a community of practice with the purpose of furthering the mathematics learned. Space does not allow a full discussion of this research, but needless to say, a sociocultural perspective continues to be an important lens for the study of mathematics teaching and learning.

Boundary Objects

The sociologists Star and Griesemer (1989) published the first definition of boundary object. According to Star and Griesemer,

Boundary objects are objects which are both plastic enough to adapt to local needs and constraints of the several parties employing them, yet robust enough to maintain a common identity across sites. They are weakly structured in common use, and become strongly structured in individual-site use. They may be abstract or concrete. They have different meanings in different social worlds but their structure is common enough to more than one world to make them recognizable means of translation. The creation and management of boundary objects is key in developing and maintaining coherence across intersecting social worlds. (p. 393)

Three particular points that make this notion of boundary object appropriate for our development of a classroom TEBO. First, the notion of a plastic object brings to mind the idea of being flexible but still maintaining the identity of the object. This is what we do with the "less is more" technology. The visuals that the students see may not be exactly the visual, analytical notion that mathematicians use, but they are recognizable and useful in both communities.

Second, boundary objects may be concrete or abstract. In the case of classroom TEBOs, we believe it is both. Seeing the actual representation of a solution or a tangent vector fields is concrete and the development of the concept of solution and how solutions are created or recovered from the differential equation or system of differential equations is abstract. For the mathematics student, this distinction is likely not as developed or complete as it is for mathematicians, but it is useful way to characterize the link that occurs between the classroom community and broader mathematical community.

Third, the "coherence across social worlds," is exactly what we mean when we refer to bridging the boundary of the mathematics classroom community and the community of mathematicians. We believe that learning is a social activity as it occurs in classrooms within institutions. Thus, although mathematics classrooms and mathematician communities may not be social worlds in the traditional definition of the term, the way they function makes it appropriate to use the concept of boundary object as being situated between the two worlds.

As previously stated, TEBOs are "software tools that adapt or extend symbolic artefacts identified from existing work practice, that are intended to act as boundary objects, for the purposes of employees' learning and enhancing workplace communication" (Hoyles, Noss, Kent, & Bakker, 2010). Described in more detail, Hoyles et al state that,

A key characteristic of a TEBO is that it involves symbolic information, typically a graph, a model expressed in algebraic symbols, or even a single numerical measure. In all the cases the meanings of the symbolic information are rendered more visible and manipulable, and therefore more accessible, by the use of interactive software tools (p. 17).

We found this further characterization of TEBOs to be a useful way to conceptualize our software tools, specifically because the symbolic information that can be used for visualization and manipulation provides a way for students to make meaning in the mathematics classroom and bridge the gap to the broader mathematical community.

Technology and Visualization in Differential Equations

Current technology has provided the computational power to develop mathematics that describes and predicts the behavior of complex real world systems that are intractable without it (Kallaher, 1999). This is extremely important in engineering and science, but teaching is better supported by the use of technology to connect algebraic and graphical representations, and aid in conceptualization and visualization. Rasmussen (2001) examined the connections students make between graphical and algebraic representations. In his study, he provided students with the autonomous differential equation $\frac{dN}{dt} = -4N\left(1-\frac{N}{3}\right)\left(1-\frac{N}{6}\right)$ and the corresponding graph of $\frac{dN}{dt}$ vs. N. He asked students in individual interviews the following questions: What is the limiting population for N(0)=2, N(0)=3, N(0)=4, and N(0)=7? Four of the six students were unable to answer the question. The typical student approach to this problem was to find answers to the first two parts by creating a sketch like that in Figure 1.

Rasmussen found that students did not view the sketch they had just created as a plot of the functions that solve the differential equation. In the words of one student, his sketch was "just a test for stability." These students had learned a graphical approach for determining stability in which the graphs they created did not carry the intended conceptual meaning. This connects with Donovan's work in that the students did not think of the graphical sketches as solution functions (Donovan, 2001). This chapter provides more insight on ways to use technology to support students' work in visualizing and then understanding about solutions to differential equations.

Habre (2000) found that students do not prefer to use graphical methods and techniques and almost always choose analytic solutions when possible. Habre's research lends further support to the claim that student conception that solutions

Figure 1. Typical student sketch for $\frac{dN}{dt} = -4N\left(1-\frac{N}{3}\right)\left(1-\frac{N}{6}\right)$

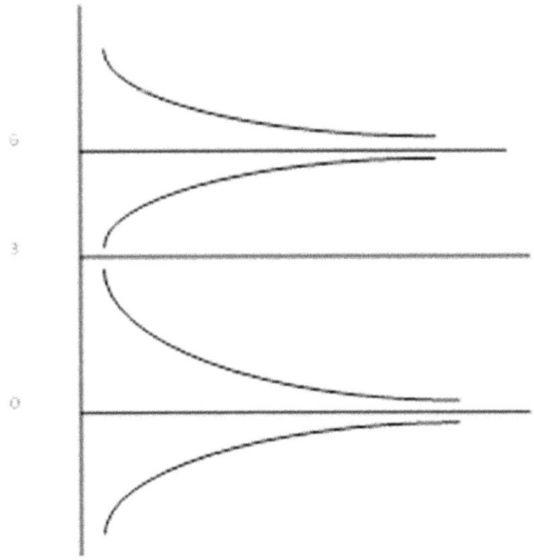

must have an analytical form are resistant to change and that moving to the graphical setting to understand differential equations is extremely difficult. Students' reluctance to value graphical solutions equally with analytic solutions was likely a result of the mathematical culture that they have experienced in many of their previous mathematics classrooms (Artique, 1992; Habre, 2000).

In recent years, more emphasis has been placed on the graphical setting and mathematics educators have suggested that more work in the graphical setting can potentially contribute to greater conceptual understanding by helping with visualization, students who are required to do more graphing may not necessarily develop better conceptualizations (Rasmussen, 2001). Research findings focusing on student understanding in reform courses indicate that graphical and qualitative approaches do not automatically translate into conceptual understanding. Students may use graphical techniques in ways that are not connected to meaning and students may have preconceived notions of what graphs look like which they utilize when looking at graphical solutions to differential equations (Rasmussen, 2001). This research on graphical understanding suggests that just adding graphical techniques to a differential equations class may not increase conceptual understanding. The learning environment and norms constituted also contribute to student understanding in the class.

There is some research, however, that graphical representations in differential equations can promote better understanding. In Habre's study (2000), for example, students used computer modules designed for specific course goals and intended to introduce students to specific concepts. Although this was not the focus of his research, Habre's interview data suggests that these modules might have been helpful to students in their development of mathematics in the graphical setting. For example, when given a vector field for the system of differential equations $x'(t) = -x+4y$, $y'(t) = -3x-y$, all the students he interviewed were able to draw an appropriate trajectory in the xy-plane and draw reasonable $x(t)$ and $y(t)$ graphs corresponding to this trajectory. Some students, however, were not able to draw the curve in three dimensions. Habre suggested that the role of such computer modules in student learning warrants further study.

In another study, Andresen (2004) reported on the use of computer algebra systems in differential equations. She studied the development of flexibility in thinking in a class studying differential equations and using the CAS Derive as an integral part of the course. Flexibility is described as a student's ability to change between perspectives and forms of representations as they reason about mathematics; in this case, she studied student's perspectives of solutions to differential equations. Her research indicated that the use of Derive was a positive influence in students' ability to develop flexibility, particularly in their movement between types of representations when solving differential equations.

The classroom TEBOS described in this chapter offer another possible way that visualization and various graphical representations can contribute to deeper understanding. In the next section we describe three such examples.

CLASSROOM TEBOS: EXAMPLES OF THE "LESS IS MORE" TECHNOLOGY

Example 1

In our first example of a classroom TEBO, students reinvent Euler's method using a Java Applet that serves as a tool for exploration and visualization. In a previous problem students worked with an unlimited resource situation that led to three outcomes: the formulation of a rate of change equation $dP/dt = kP$, the emergence of the associated tangent vector field (see Rasmussen and Marrongelle (2008) for details on how the vector field arises out of student thinking), and the introduction of separation of variables to determine the exact solution functions. Building on this previous work, students modify the unlimited resource scenario to the more realistic limited resource situation. This modification leads to differential equations such as $\frac{dP}{dt} = 0.3P(1 - \frac{P}{10})$. This differential equation is separable, but carrying out the technique involves the method of partial fractions and hence can be analytically cumbersome, without any clear benefit for developing new ideas. Thus, instead of pursuing solutions to this differential equation using the technique of separation of variables, students are invited to explore graphical and numerical approaches using the applet called DEVectors.

As shown in Figure 2, the DEVectors applet displays a grid of vectors whose slopes are determined by the differential equation. In designing the applet we had to make a choice as to where to place the vector that corresponds to the slope at any particular point in the plane. Should the vector be centered on the point? Should the tail of the vector be associated with the slope at each point? Should the vector have an arrow at its tip or not?

Because we wanted to foster the imagery of starting at a particular point and traveling along a straight line path for some amount of time, we chose to associate the tail of the vector with the point at which the derivative is computed. We also chose to include an arrow at the tip of the vector in order to suggest motion or movement that would lead to tracing out a curve. Another design decision was whether to make the vectors uniform in length or not. In order to promote reinvention of Euler's method where the change is the independent variable (in this case time, *t*)

Figure 2. DEVectors applet created for $\frac{dP}{dt} = 0.3P(1 - \frac{P}{10})$

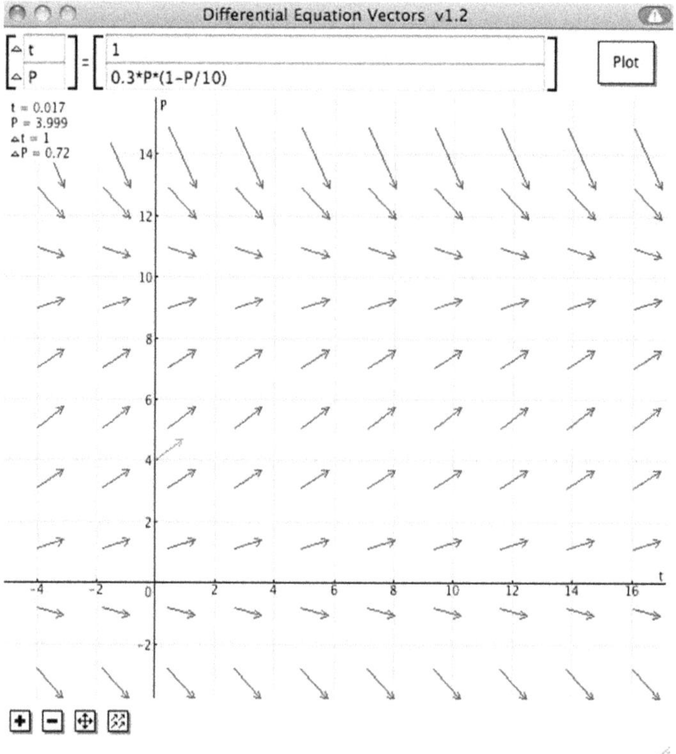

is held constant, we decided to take an unconventional route and plot vectors such that the base is of uniform length but whose heights are not; smaller slope values have vectors that are slightly smaller in length than vectors with larger slope values.

Placing the cursor over any point in the plane displays a red tangent vector associated with that point. An example of such a red vector is shown in Figure 2 near the point (0, 4). The students can move cursor (and hence the vector) about in the plane and its slope changes accordingly. Thus, students can kinesthetically experience the emergence of a solution by "following the slope field". The applet does not sketch any curves in for the student, like software tools for experts, but rather students must do this using their imagination and by hand sketching. The applet also allows the user to move about the coordinate axes and the user can zoom in and out. The number of displayed vectors can be increased or decreased, making the visual display more or less dense.

In the following task students use DEVectors to work in pairs to graphically and numerically create an approximate solution to the given differential equation.

Use the Java applet DEVectors to create a tangent vector field for the rate of change equation $\frac{dP}{dt} = 0.3P(1 - \frac{P}{10})$. For each of the problems next, place a transparency on your computer screen and use a marker to actually draw on the transparency, leaving a record of where you've been as you move a point around in the plane. Work in pairs to do this: one person works the mouse and the other person uses the marker to trace the path on the transparency.

Figure out a way to link the tangent vectors together to create a graph of the population versus time if at time 0 we know there are 5 fish in the lake (think of 10 as scaled for say, 1000 or 10,000). Create a table of (t, P) values to match your graph.

Students work in pairs using the DEVectors Java Applet. One student places a transparency sheet on the computer screen and uses a marker to trace out the path of the solution. The other student uses the mouse to move the red highlighted vector in a tip to tail manner. As the arrow moves, the first student draws the connecting vectors on the transparency. This part of the instructional task not only provides visualization of solutions to a differential equation, it also promotes the use of student's tactile sense to experience a solution. We believe that use of all the senses promotes an embodied view of learning (Lakoff & Nunez, 2002) and this is helpful as students begin to grasp how the solution to a differential equation is defined by the differential equation itself.

Because students have to create a table of *(t, P)* values corresponding to their tip to tail construction, they either have to rely on the numerical values show in the top left corner of the applet (See Figure 2) or they have to devise a computational approach for themselves. Students readily realize that their ability to accurately place vectors by hand and to read off the (t, P) from the applet is error prone and in need of a more systematic method. As the instructional sequence continues students are given the following two tasks that culminate in the reinvention of Euler's method.

Without using the java applet, create a table of (t, y) values that approximates the solution to the differential equation dy/dt = 0.1y (t – y), with initial condition y (0) = 4. Check your results with the applet and provide a graph of your approximation.

Generalizing your result. Create an equation-based procedure/algorithm that would allow you generate a table of values for any differential equation and any given initial condition.

Key to success on these tasks is realizing that the slope is a ratio of two covarying quantities. While this may seem obvious to those of us who already know Euler's method, students need time, and the opportunities afforded by the DEVector applet, to create an algorithm that leverages thinking about slope as a ratio. In such, the

DEVector applet functions as a boundary object between the classroom community and the broader mathematical community. As students use the applet and develop their algorithms, they reinvent important mathematics and become increasingly enculturated in the discipline of mathematics.

Example 2

The second example occurs a few weeks later after students have learned various analytic techniques for solving first order differential equations. In the next example, we use the same DEVector applet to conduct an investigation that leads students to develop strong intuition for conditions related to the uniqueness theorem. The uniqueness theorem is usually presented as part of a section on "Existence and Uniqueness of Solutions" in a differential equation course. The presentation is very symbolic and students rarely have any opportunity to explore conditions under which solutions are unique. Instead, they are usually presented with the theorems as a finished notion. Theoretically, one version of the theorem states that if the derivative with respect to the dependent variable of the differential equation is bounded on an interval, the solutions in that region are unique. In other words, the solution functions never touch or cross. The task we use supports a way for students to experience this condition first in a visual and experiential method, with the instructor subsequently providing the formal statement of the theorem (Rasmussen & Ruan, 2008).

Students use the tangent vector field applet to investigate how a solution to $dh/dt=-h$ and $dh/dt=-h^{1/3}$ creates different possible imagined paths for a helicopter landing. Next is the sequence of tasks related to these two different differential equations.

A group of scientists at the FAA has proposed the rate of change equation $\frac{dh}{dt}=-h$ as a means to predict the height of a helicopter as it nears the ground while a different group of scientists proposed the rate of change equation $\frac{dh}{dt}=-h^{\frac{1}{3}}$ (in both rate of change equations h is in feet and t is in minutes). The scientists of course want their models to predict that a helicopter actually lands – but does either or both of the proposed models predict this? The purpose of the following problems is to examine this question.

1. Getting familiar with the differential equations.
   Just by examining the proposed rate of change equations, what can you say about the height of the helicopter as predicted by $\frac{dh}{dt}=-h$ and by $\frac{dh}{dt}=-h^{\frac{1}{3}}$?
   Explain your ideas and sketch a graph of height versus time.

What does each of the proposed rates of change equations say about the solution to the differential equation if the helicopter is already on the ground? Explain your ideas and sketch a graph of height versus time.

2. When, if ever, is this helicopter going to land?
   After generating both vector fields, click on a point in the plane so that a tangent vector for the rate of change equation $\frac{dh}{dt} = -h$ is activated and move the point with its associated tangent vector around in the plane so as to create a graph of the solution with initial condition $h = 2$ at $t = 0$ (think of 2 as 200 or 2000 feet above ground). Do the same thing for the rate of change equation $\frac{dh}{dt} = -h^{\frac{1}{3}}$. Do both, none, or just one of the graphs appear to eventually touch the constant solution function $h(t) = 0$? Explain your ideas.
3. The manner in which the tangent vectors changed as h approaches 0 suggested to one group of students that one of the differential equations predict that the helicopter will land in a finite amount of time while the other differential equation predicts that the helicopter will never actually touch ground. Figure out a way to use the technique of separation of variables to determine the validity of this conjecture for a helicopter with initial condition $h(0) = 1$.
4. In the task previously you found that the solutions to $\frac{dh}{dt} = -h$ never touch the equilibrium solution h(t) =0 whereas the solutions to $\frac{dh}{dt} = -h^{\frac{1}{3}}$ do touch the equilibrium solution h(t) = 0. Why, geometrically, can you explain this finding? Here are two possible ways to explore *why* this is the case.
   a. Click on a tangent vector for $\frac{dh}{dt} = -h$ and drag it directly down (i.e., don't move left or right just straight down toward zero). What do you notice about the *rate of change of the tangent vector* as you get close to $h = 0$ and what, if anything, does this suggest about whether the differential equation predicts a landing in finite time? Repeat for $\frac{dh}{dt} = -h^{\frac{1}{3}}$.
   b. Sketch graphs of dh/dt versus h for both differential equations and explain what the slope of this graph at h = 0 says about the way the solution graphs (that is, the h(t) graphs) approach the constant function h(t) = 0.

The students begin this task by using DEVectors to consider the solution paths of the two differential equations $\frac{dh}{dt} = -h$ and $\frac{dh}{dt} = -h^{\frac{1}{3}}$. They can visualize the two possible solutions by following the TVF on the applet, and for most students, the paths look very nearly the same (See Figure 3).

However, on closer investigation, one or more pairs of students typically sense that the right TVF looks like the solution path would allow the helicopter to land, when the left one does not. They reason this way: As the tangent vectors on the left seem to get flatter and flatter and hence the solution graph never reaches the *t*-axis while the one on the right shows vectors that are indicative of a graph that touches the *t*-axis. (As an aside, the program does not show tangent vectors below the t-axis in the right TVF, but this is not part of the model, and the instructor can note this). Much earlier, students were introduced to equilibrium solutions, or constant solutions that solve the differential equations. The solution *h(t)=0* is an equilibrium solution for both the differential equations and students can see this as well from the differential equations and the TVFs. The issue then is this: If the solution actually touches the t-axis, then at that point, two solutions touch (the decreasing solution and equilibrium solution) and so the solutions are not unique. This motivates the need to figure out how to know when solutions will or will not be unique. Task 4 offers students two different ways to explore this issue.

Further work in the class encourages students to "zoom in on" the tangent vector field and pull the red movable vector straight down toward the *t*-axis (and solution *h(t)=0*). At some point, some of the students notice that the change in the slope of the tangent vector as *h* approaches 0 is gradually approaching 0 for the $\frac{dh}{dt} = -h$ and that the tangent vector for the $\frac{dh}{dt} = -h^{\frac{1}{3}}$ is getting less steep, but at some point

Figure 3. TVFs for dh/dt=-h and dh/dt=-h$^{1/3}$

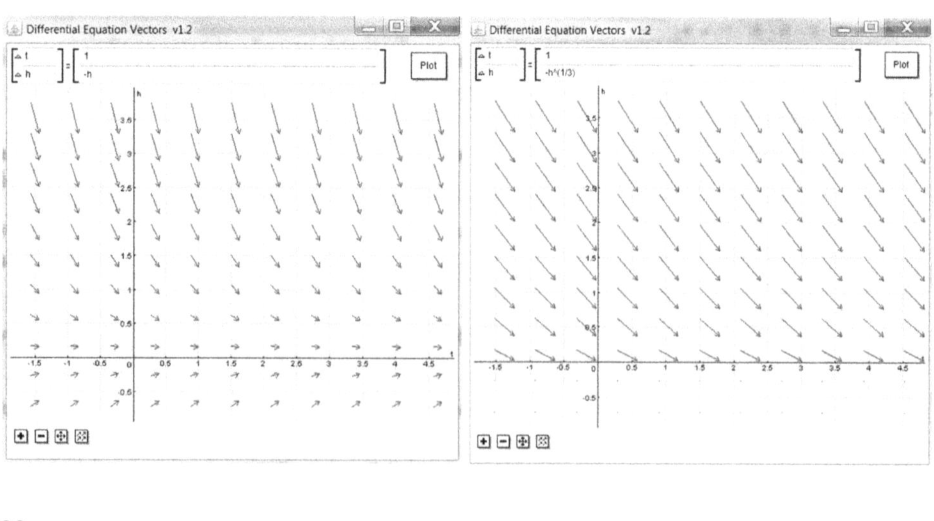

it "snaps" or goes abruptly to a slope of zero. This visualization leads to the following hypothesis about when solutions do or do not touch each other (that is, they are unique).

The rate of change (or derivative) of the differential equation with respect to *h* (the dependent variable) is the concept that the students are visualizing in this classroom TEBO shown in Figure 4. When the tangent vector (representing the derivative) of the second solution "snaps", it shows the students visually how the derivative of the derivative (rate of change) is infinitely large or unbounded near h= 0. From an expert's perspective, this is related to the stronger Lipschitz condition and leads to a version of the uniqueness theorem in that if the derivative is unbounded, there is no guarantee of uniqueness in the region. Seeing it visually first is very powerful and helps make the conditions for the uniqueness theorem sensible and meaningful. Anecdotal evidence indicates that students are now more capable and proficient in using the uniqueness theorem as a tool for reasoning about solutions in subsequent lessons. Next is the version of the theorem that students eventually reason with and use.

Figure 4. Zoomed in view of the TVF for dh/dt=-h$^{1/3}$

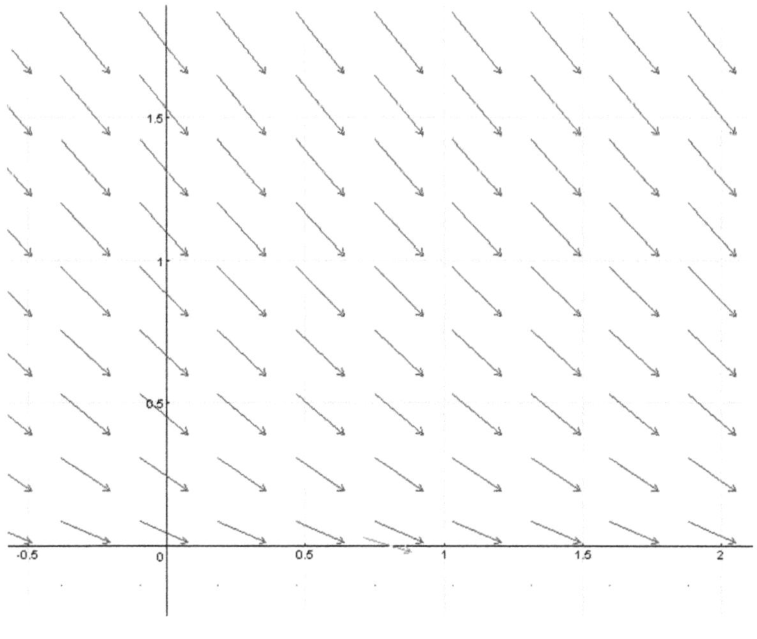

The Uniqueness Theorem

More often than not it is impossible to actually find the exact solution to a differential equation and therefore this approach (solving analytically) for determining if solutions ever actually touch an equilibrium solution function is not feasible. The uniqueness theorem offers a theoretical way to determine if solutions to a differential equation ever actually touch an equilibrium solution function.

For a differential equation of the form $\frac{dy}{dt} = f\left(t, y\right)$, the uniqueness theorem can be stated in the following way:

If f(t,y) is continuous on $a \leq t \leq b$, $c \leq y \leq d$, and there exists a constant L such that

$|f(t,y) - f(t,z)| \leq L|y - z|$ for all $t \in [a, b]$, and all $y, z \in [c, d]$,

then the solution to $dy/dt = f(t, y)$ with $y(t_0) = y_0$ where $t_0 \in [a, b]$ and $y_0 \in [c, d]$ is unique for all $t \in [a, b]$ such that $y(t) \in [c, d]$.

Example 3

There are different ways to conceptualize the solution to a system of two first order differential equations, *dx/dt = f(x,y)* and *dy/dt = g(x,y)*. One way is to think of a solution as a pair of functions, *(x(t), y(t))* that satisfy the differential equations. Another way to conceptualize a solution, particularly in the realm of time-based dynamical systems, is that a solution is a curve in *(x, y, t)* space. Such a 3-D curve then can be analytically expressed by two functions, *x(t)* and *y(t)*, or as a parameterized curve in the *(x, y)* plane. As our example illustrates, first thinking of a solution as a 3-D curve allows students to move from an informal, imaginary situation to more formal analytical understanding of mathematics that brings coherence to the various ways of conceptualizing a solution to a system of two differential equations.

Students begin the unit on systems of differential equations with a realistic but somewhat imaginary activity. They use flexible wire spirals or fabric covered wires (pipe cleaners) to model a red trace that is imagined to be released from a propeller on an airplane taxiing in a straight line on a tarmac. The airplane is observed from four different vantage points. The activity is created by the following task sheet:

Three Dimensional Visualization

A crop duster plane with a two blade propeller is taxiing down a runway. On the end of one of the propeller blades, which are rotating clockwise at a slow constant speed, is a noticeable red paint mark. Imagine that for the first several rotations of the propeller blades the red mark leaves a "trace" in the air as the plane makes it way down the runway.

1. Using a red cube as the paint smear, simulate this scenario and create the trace with a pipe cleaner or slinky. Sketch the *ideal perspective* for persons Alpha, Beta, Gamma, and Delta. Alpha's position gives him a 3-D view of the trace, while the positions of Gamma, Beta, and Delta all provide a 2-D view of the trace. What information, if any, is lost in the various perspectives?
2. Sketch your ideas for each of the following:
   - What if there was another paint mark on other end of the propeller, what, *ideally*, do the four observers see then? How does the trace of this mark relate to the previous trace?
   - What if there was a paint mark on the center of the propeller blade mechanism. What do the observers ideally see then?
   - How ideally would each observer see all of the previous paint marks simultaneously?
   - What do the four observers ideally see if the red mark slowly drifts toward the center?
   - What do the four observers ideally see if the propeller is <u>not</u> rotating and the red mark drifts away from the center at a rate proportional to its distance from the center as the plane taxis down the runway?

This task creates an opportunity for students to visualize how the three dimension curve of a spiral will appear from different perspectives. Although students only spend half a class period on this task, they will often refer to the "airplane task" when they want to express their thinking about the solutions to the systems. In the task as asked in part A, Person Alpha sees a spiral in 3-D, Person Beta and Delta see a curve similar to a sine or cosine curve (but with a phase shift of each other) and person Gamma sees a circle (if depth perspective is ignored). After considering alternative scenarios as listed previously, the students connect this visualization to a new Java Applet called DEExplorer.

In the classroom TEBO called DEExplorer, students use an applet that allows the user to enter a system of two differential equations and an initial condition. The applet then generates a table of *(t, x, y)* values and a corresponding curve in 3-D. The table of values is generated with either the Euler or Runge Kutta method. Students can then click on the 3-D curve and move it around, in a continuous manner, to show different perspectives of the 3-D curve. In particular, they can see what the curve looks like in the x-t, y-t, and x-y planes. These different planes connect directly back to the "airplane task" and the Alpha, Beta, Gamma, and Delta views (See Figure 6 for three different views created for the system).

Students often call the three-dimensional solution curve (See Figure 5a) the "mother curve" because they can use the cursor to manipulate the curve to see the many different views and how they are all related. They can also change the initial condition and have multiple solution curves plotted simultaneously. These options afford students significant investigative power. Through their work on this classroom TEBO, students develop an understanding of what it means to be a solution to a system and the various ways to represent solutions, thus connecting them to the mathematician community of practice. It is interesting that experts in the mathematical community often conceptualize solutions to the basic predator-prey model we see in this first instructional task as oscillatory curves. When mathematicians see this classroom TEBO, they are able to enter into the classroom mathematics community from the learner's perspective.

One additional feature of the DEExplorer that is noteworthy is that students see the 3-D curve generated in real time as opposed to appearing all at once. Because time-based dynamical systems are more and more important in differential equations, seeing these curves actually be generated in real time adds another perspective that is powerful for the students. Thus, because of this feature and the ability to physically change the view, this applet serves as a classroom TEBO for the students to develop mathematical understanding about solutions to systems and how they appear in different planes and in 3-space. In addition, this TEBO motivates the need for the phase plane and vector fields, which are the tools of differential equation and dynamical system experts.

From an expert's point of view, the phase plane is one of the most useful ways to visualize solutions to autonomous systems of two differential equations, yet from the learner's perspective the motivation and origins of the phase plane are often not

Figure 5. Three dimensional visualization task

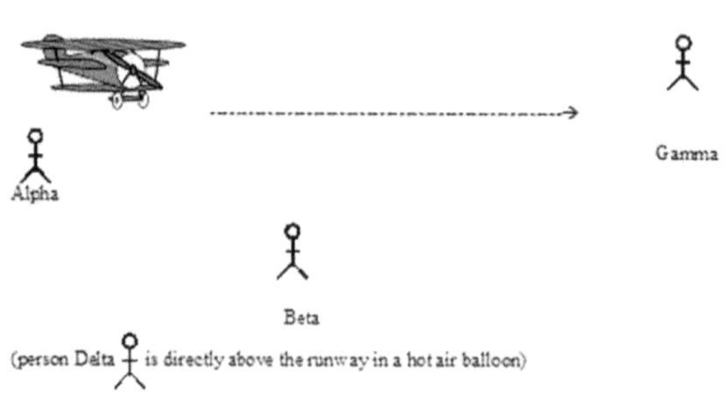

Figure 6. Three views that can be represented with the DEExplorer applet (a) 3-D view of solution (b) x vs. t view of the 3-D solution (c) x vs. y view of the 3-D solution

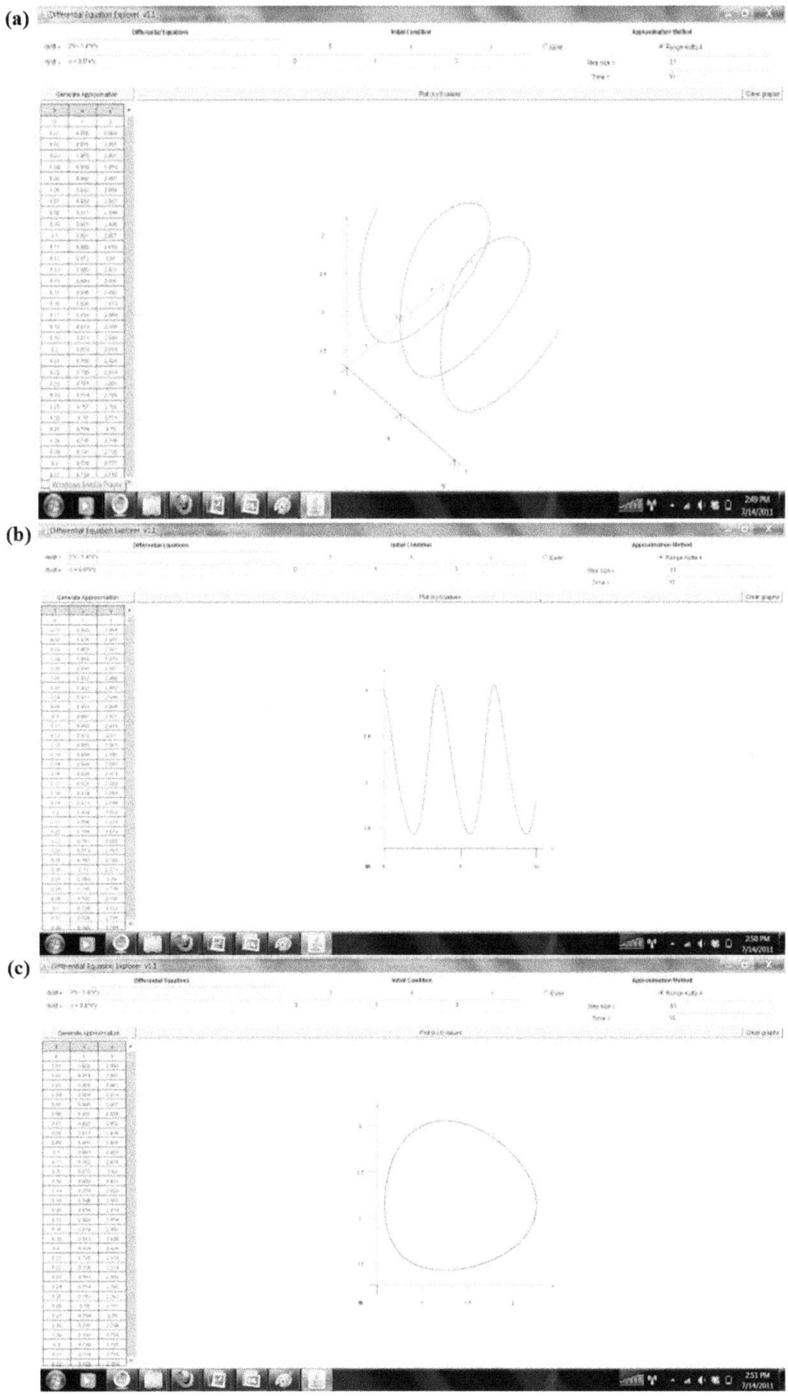

well understood. The classroom TEBO DEExplorer provides an opportunity for students to first build coherent imagery for the different ways to visualize that gives rise to the phase plane and situates it within the multiple ways to visualize solutions. For example, one mathematics instructor remarked, "This is a wonderful applet. I can see how this can help students understand the solution to a system. I especially like the feature that the solution curve can be viewed from many different angles. My students used this tool for the rest of the semester after they were given it, even when they had learned the analytical solution method, to help with their thinking."

CONCLUSION

The two classroom TEBOs, the Java Applets called DEVector and DEExplorer, serve as boundary objects between the classroom mathematics community and the broader community of mathematicians. In each of the three examples, the students use technology to visualize, develop intuitions, and reinvent important mathematics. The graphs they engage with are animated and interactive, which provides ways for them to understand important ideas. In all cases, the graphical representations serve as a connector between the students who are in the classroom and the broader mathematical community. By design, the classroom TEBOs do not give students all of the information that an expert would require, but instead require them to participate in exploration and experimentation in order to develop insight and to motivate the need for all the information that an expert would like to see.

In this chapter, we offered two boundary objects that are visualization tools for students in undergraduate mathematics. We adapted the notion of a TEBO, a technology-enhanced mathematical boundary object, to the mathematics classroom as a way for mathematics educators and mathematicians to describe and analyze technology tools at the university level. Because the tools are specifically not designed to do everything required by an expert, they play the role of boundary objects. Students investigate solutions to differential equations in an interactive environment, and their active involvement helps them to develop intuition and understanding.

We note that much research has shown that visualization and conceptual understanding are inextricably linked for students as they learn mathematics (i.e., Presmeg, 1997; Arcavi, 1999; Duval, 1999; Nardi & Ianone, 2003). This is why it is important to ground students' new mathematical understanding in visualization when possible. We contend that experts may not need this same visual view of concepts and thus use software programs that do much more of the mathematics. The "less is more" applets detailed in this chapter do less of the mathematics for students and thus students must do more of the mathematical thinking.

Sometimes Less is More

The three examples we provided show how the TEBOs are special in the way they introduce the students into the differential equations learning community. In the first example, the students, in the early stages of their instruction in solving ODEs, experiment with tangent vector fields and see how the slopes of the tangents as determined by the differential equation provide a kind of roadmap for the solution to follow as it emerges. This supports their reinvention of a numerical technique; students continue to return to the technology and interweave their understanding of the visual with the analytical as the course continues. The second example shows how students use the DEVector applet to develop an intuitive and later formal understanding of uniqueness. The capability of the tool for the students to move the vectors and thus study carefully the slope change as the dependent variable changes is very helpful in building understanding of the theorem.

In the third example, the students connect an imaginary experience, making airplane propeller traces, to the DEExplorer applet to model a three dimensional representation of a solution to a system of two differential equations. The view of analytic solutions as two functions emerges from their looking at the 3-D "mother curve" and rotating it in ways that then show the x-t and y-t views, which are representation of the two analytic function solutions. Mathematics instructors consistently find that using this program benefits students even after they have been provided with more complex, expert software.

As research from the Dienes' blocks model showed, mathematical tools are sometimes only valuable when the conception is already present for a student, and thus the tools do not serve the purpose that is intended (Dienes, 1960). With the classroom TEBOs in this chapter, "less is more" and students can actually think about the new mathematics in a simple and powerful visual environment. This provides support for them as they move into analytical and numerical analysis of solutions of differential equations and interweaves with this continued learning. Possibly more of these TEBOs can be developed in other mathematics courses at both the secondary and tertiary classrooms.

We do not claim in this chapter that these should be the only visualization tools used in a mathematics classroom. Other tasks, teacher input, student engagement in justification and investigation work together to provide an environment that will allow students to develop their own ways of thinking about mathematics that connect with conventional and formal understandings. The ways in which classroom mathematics and expert mathematics can be connected and bridged by TEBOs is an area that is in need of further research. Studies are also needed that compare student learning of concepts and procedures in classrooms that use TEBOs to those classrooms that use expert tools and to classrooms that use little to no technology. Our previous comparison studies, while not focused on the effect of technology per se, suggest that students in problem-based, inquiry oriented classrooms learn concepts better and retain this knowledge longer than students in traditionally taught classrooms (Kwon, Rasmussen, & Allen, 2005; Rasmussen, Kwon, Marrongelle, Allen, & Burtch, 2006). These studies also indicate that there is no significant difference in students' proficiency with procedural proficiency in solving differential equations without any technology. More work specifically focused on the role of technology and visualization, however, is needed.

REFERENCES

Arcavi, A. (1999). The role of visual representations in the learning of mathematics. In F. Hitt & M. Santos (Eds.), *Proceedings of the 21st North American PME Conference,* (vol. 1, pp. 55-80). PME.

Boaler, J. (2002). The development of disciplinary relationships: Knowledge, practice and identity in mathematics classrooms. *For the Learning of Mathematics, 22*(1), 42–47.

Bowker, G. C., & Star, S. L. (1999). *Sorting things out: Classification and its consequences.* Cambridge, MA: MIT Press.

Burton, L. (1999). The practices of mathematicians: What do they tell us about coming to know mathematics? *Educational Studies in Mathematics, 31*, 127–143.

Cobb, P., & Yackel, E. (1996). Constructivist, emergent, and sociocultural perspectives in the context of developmental research. *Educational Psychologist, 31*(3/4), 175–190.

Dienes, Z. (1960). *Building up mathematics* (4th ed.). London, UK: Hutchinson Educational Ltd..

Duval, R. (1999). Representation, vision and visualization: Cognitive functions in mathematical thinking – Basic issues for learning. In F. Hitt & M. Santos (Eds.), *Proceedings of the 21st North American PME Conference,* (vol. 1, pp. 3-26). PME.

Goos, M. (2004). Learning mathematics in a classroom community of inquiry. *Journal for Research in Mathematics Education, 35*, 258–291. doi:10.2307/30034810.

Gravemeijer, K. (1994). *Developing realistic mathematics education.* Utrecht, The Netherlands: CD-β Press.

Habre, S. (2000). Exploring students' strategies to solve ordinary differential equations in a reformed setting. *The Journal of Mathematical Behavior, 18*, 455–472. doi:10.1016/S0732-3123(00)00024-9.

Herzig, A. (2004). Becoming mathematicians: Women and students of color choosing and leaving doctoral mathematics. *Review of Educational Research*, *74*(2), 171–214. doi:10.3102/00346543074002171.

Hoyles, C., Noss, R., Kent, P., & Bakker, A. (2010). *Improving mathematics at work: The need for techno-mathematical literacies*. New York: Routledge.

Kwon, O. N., Rasmussen, C., & Allen, K. (2005). Students' retention of knowledge and skills in differential equations. *School Science and Mathematics*, *105*(5), 227–239. doi:10.1111/j.1949-8594.2005.tb18163.x.

Lakoff, G., & Nunez, R. (2000). *Where mathematics comes from*. New York, NY: Basic Books.

Lambertus, A. J. (2010). *Characterizing gender diverse graduate mathematics departments as communities of practice*. Doctoral dissertation.

Lave, J., & Wenger, E. (1991). *Situated learning: Legitimate peripheral participation*. Cambridge, UK: Cambridge University Press. doi:10.1017/CBO9780511815355.

Lee, C. P. (2007). Boundary negotiating artifacts: Unbinding the routine of boundary objects and embracing chaos in collaborative work. *Computer Supported Cooperative Work*, *16*, 307–338. doi:10.1007/s10606-007-9044-5.

Lesh, R., Post, T., & Behr, M. (1987). Dienes revisited: Multiple embodiments in computer environments. In Wirsup, I., & Streit, R. (Eds.), *Development in School Mathematics Education Around the World* (pp. 647–680). Reston, VA: National Council of Teachers of Mathematics.

Nardi, E., & Iannone, P. (2003). Mathematicians on concept image construction: Single 'landscape' vs 'your own tailor-made brain version'. In N. Pateman, B. J. Dougherty, & J. Zillox (Eds.), *Proceedings of the 27th PME International Conference*, (vol. 3, pp. 365-372). PME.

Pratt, N., & Kelly, P. (n.d.). *Mapping mathematical communities: Classrooms, research communities and master classes*. Retrieved July 3, 2011 from http://orgs.man.ac.uk/projects/include/experiment/kelly_pratt.pdf

Presmeg, N. C. (1997). Generalization using imagery in mathematics. In English, L. D. (Ed.), *Mathematical reasoning: Analogies, metaphors and images* (pp. 299–312). Mahwah, NJ: Erlbaum.

Rasmussen, C., & Kwon, O. (2007). An inquiry oriented approach to undergraduate mathematics. *The Journal of Mathematical Behavior*, *26*, 189–194. doi:10.1016/j.jmathb.2007.10.001.

Rasmussen, C., Kwon, O., Allen, K., Marrongelle, K., & Burtch, M. (2006). Capitalizing on advances in mathematics and K-12 mathematics education in undergraduate mathematics: An inquiry-oriented approach to differential equations. *Asia Pacific Education Review, 7*, 85–93. doi:10.1007/BF03036787.

Rasmussen, C., & Marrongelle, K. (2006). Pedagogical content tools: Integrating student reasoning and mathematics in instruction. *Journal for Research in Mathematics Education, 37*, 388–420.

Rasmussen, C., & Ruan, W. (2008). Using theorems-as-tools: A case study of the uniqueness theorem. In Carlson, M., & Rasmussen, C. (Eds.), *Making the Connection: Research and Teaching in Undergraduate Mathematics*. Washington, DC: The Mathematical Association of America.

Rasmussen, C., Zandieh, M., King, K., & Teppo, A. (2005). Advancing mathematical activity: A view of advanced mathematical thinking. *Mathematical Thinking and Learning, 7*, 51–73. doi:10.1207/s15327833mtl0701_4.

Rasmussen, C., Zandieh, M., & Wawro, M. (2009). How do you know which way the arrows go? The emergence and brokering of a classroom mathematics practice. In Roth, W.-M. (Ed.), *Mathematical representations at the interface of the body and culture* (pp. 171–218). Charlotte, NC: Information Age Publishing.

Solomon, Y. (2007). Not belonging? What makes a functional learner identity in undergraduate mathematics? *Studies in Higher Education, 32*(1), 79–96. doi:10.1080/03075070601099473.

Star, S. L., & Griesemer, J. R. (1989). Institutional ecology, translations and boundary objects: Amateurs and professionals in Berkeley's museum of vertebrate zoology, 1907-1939. *Social Studies of Science, 19*(3), 387–420. doi:10.1177/030631289019003001.

Stars, L., & Griesemer, J. R. (1989). Institutional ecology, translations and boundary objects: Amateurs and professionals in Berkeley's museum of vertebrate zoology, 1907-1939. *Social Studies of Science, 19*(3), 387–420. doi:10.1177/030631289019003001.

Van Oers, B. (2001). Educational forms of initiation in mathematical culture. *Educational Studies in Mathematics, 46*, 59–85. doi:10.1023/A:1014031507535.

Wenger, E. (1998). *Communities of practice: Learning, meaning, and identity*. Cambridge, UK: Cambridge University Press.

Chapter 3
"Click, Drag, Think!" Posing and Exploring Conjectures with Dynamic Geometry Software

Thomas Gawlick
Leibniz Universität Hannover, Germany

ABSTRACT

The author points out that to fully exploit the heuristic potential of Dynamic Geometry Software (DGS) and to increase the heuristic literacy of students, extant DGS teaching units have to be ameliorated in several ways. Thus the author develops a twofold conceptual framework: heuristic reconstruction and heuristic instrumentation of problems. Its origin is rooted in the literature, its use is demonstrated by various examples, and its value is made plausible by an introductory teaching unit an advanced case study.

OVERVIEW

DGS has become an established tool in the mathematics classroom. Though its heuristic value is often stressed, the literature on DGS usage reveals some shortcomings in its existing use. To see what is missing, we first reconsider models of problem solving and proving by Polya and Boero. With these theoretical tools, we can state our main idea: to further success in problem solving, a *heuristic reconstruction* of tasks is suitable, which we design analogously to Polya's problem solving scheme. To support learners, the *heuristic instrumentation* of problems has also to be considered, especially for best practice in utilizing DGS. We develop a framework for

DOI: 10.4018/978-1-4666-4050-4.ch003

this by refining Arzarello's list of *dragging modalities*. Both schemes are illustrated by detailed examples from teaching material of our ongoing research on DGS based problem solving. Readers that are more interested in concrete suggestions for DGS-based teaching may skip the theoretical framework and start the subsection "Heuristic Reconstruction of a Concrete Example," using the material before for later reference if necessary.

We also sketch an ongoing empirical study that will serve to evaluate our approach. Meanwhile, we point out its possible merits by an advanced case study that illustrates how far learners' heuristic abilities may evolve when heuristic strategies are properly instrumented.

THEORETICAL BACKGROUND

Dynamic Geometry Software as a Heuristic Tool

Dynamic Geometry Software (DGS) is widely recognized as a tool of visualization that may further students' progress; see Laborde & Laborde (1995). From the beginning its heuristic role was stressed: "The changes in the solving process brought by the dynamic possibilities of Cabri come from an active and reasoning visualization, from what we call an interactive process between inductive and deductive reasoning" (Laborde & Laborde 1992). Teaching material that draws on this capability was subsequently developed by Elschenbroich (1997). Meanwhile, the literature contains a plethora of proposals for the utilization of DGS even in nongeometric situations (e.g. Gawlick 2003).

However, in a critical review of the literature, Hölzl (2001) concludes a closer look leads one to question whether the software is really used in a methodical and activating way of knowledge acquisition. Rather, it seems that DGS is used in a solely verifying manner in that the learners need only vary more or less prefabricated geometric configurations to confirm empirically more or less explicitly stated conjectures (e.g. that the altitudes in a triangle always meet in one point). In contrast, Hölzl reports on the long-term application of DGS as an integral part of the learning environment that DGS possesses a considerable heuristic potential, especially for transformation geometry. Yet, the application of DGS requires thorough consideration: "Dynamics per se is not a didactical advantage" (Hölzl, 1999). It will be most favorable where an objective instrumental requirement meets advanced mathematical experience. Students were observed using the drag mode as a graphic tool, as test mode, or as a search mode–but the use as a test mode remained shaky even after two years.

Hölzl's results clearly show the need to enhance students' capability to make heuristic use of DGS: His students make only a limited use of the drag mode, up to the point where students pose nontrivial conjectures only from static drawings, even without attempting to verify them by dragging–though the drag mode as test mode was part of Hölzl's schedule!

Analogous phenomena have been observed for CAS usage by Artigue (2002): "Our attention was attracted by the slowness and windings of this instrumental genesis ... The economic strategies of use of the TI-92 were rarely chosen." She utilizes the theoretical framework of the instrumental approach (Verillon & Rabardel, 1995): "The instrument is differentiated from the object, material or symbolic, on which it is based and for which they use the term "artefact." It is a mixed entity, constituted for the one part of an artefact, and for the other, of schemes which make it an instrument. … through a process, called *instrumental genesis* .... In the first direction, instrumental genesis is directed towards the artefact, loading it progressively with potentialities, and eventually transforming it for specific uses; this is called the *instrumentalisation* of the artefact. In the second direction, instrumental genesis is directed towards the subject, and leads to the development or appropriation of schemes of instrumented action which progressively constitute into techniques that permit an effective response to given tasks. This latter is what is properly called *instrumentation*." Thus we can conceptually summarize Hölzl's findings as follows: The instrumentation of utilization schemes for DGS requires deliberate efforts over an extended period of time. To that end, we later draw on the dragging modalities of Arzarello (1998) as starting point for a theoretical framework.

Research on Problem Solving

Polya (1945) suggests mathematical problem solving to be performed in four main steps, namely:

1. Understanding the Problem
2. Devising a Plan
3. Carrying out the Plan
4. Looking Back

Polya aims at guiding the problem solver through the solution process by a sequence of questions for each step (see next). Polya's steps are not necessarily apt to describe students' actual behavior–namely, psychological research like Newell and Simon (1972) does not support the natural occurrence of "devising a plan", and the literature is inconclusive about its benefit: Forehand (1967) suggests that it might be helpful while Hayes (1966) indicates the contrary. In this vein, Kilpatrick (1967) asserts "that whatever merits Polya's list has for teaching problem solving, it is of limited usefulness, as it stands, for characterizing the behavior of these subjects. … Subjects seemingly did not exhibit behavior even remotely resembling actions suggested by the heuristic questions. For example, no subjects asked themselves aloud whether they were using all of the essential notions of the problem."

However, Schoenfeld (1985) parsed problem solving protocols into six episodes (consistent behavior of the problem solvers) that can be interrelated to Polya's phases (Rott, 2011). Schoenfeld (1992) summarizes his results: "Approximately 60% of the protocols (without prior training) were of the type ..., where the students read the problem, picked a solution direction (often with little analysis or rationalization), and then pursued that approach until they ran out of time. In contrast, successful solution attempts came in a variety of shapes and sizes." This clearly shows that the heuristic (and metacognitive) literacy of students deserves some improvement. Schoenfeld (1985) reports on successful training by breaking down several general heuristics to concrete strategies that are applicable to a narrower, but more easily identifiable class of problems. Schoenfeld's results could be replicated with more subjects and a no-treatment-control group by Rott & Gawlick (submitted).

Research on Theorem Proving

Theorem proving may be viewed as the topmost sub discipline of mathematical problem solving. For the teaching and learning of proof it is necessary to guide students to master the process of proof (and not only to deal with finished proofs). Boero (1999) thus proposed an expert model of the proof process that comprises the following phases:

1. Production of a conjecture (including exploration of the problem situation, identification of "regularities", identification of conditions under which such regularities take place, identification of arguments for the plausibility of the produced conjecture, etc.).
2. Formulation of the statement according to shared textual conventions.
3. Exploration of the content (and limits of validity) of the conjecture; heuristic, semantic (or even formal) elaborations about the links between hypotheses and thesis; identification of appropriate arguments for validation, related to the reference theory, and envisaging of possible links amongst them.
4. Selection and enchaining of coherent, theoretical arguments into a deductive chain, frequently under the guidance of analogy or in appropriate, specific cases, etc.

5. Organization of the enchained arguments into a proof that is acceptable according to current mathematical standards.
6. Approaching a formal proof (This phase may be lacking in mathematicians' theorems).

The virtue of this model will become clear when we use it as rationale in developing teaching units for theorem proving and problem solving.

ONGOING RESEARCH ON HEURISTIC LITERACY

In our research on the influence of heuristic strategies on problem solving, we thus want to further address the questions

1. What heuristics are present in students' problem solving processes?
2. What heuristics are present in students' problem solving processes?
3. Are heuristics helpful to activate problem solvers' previous knowledge?
4. To what extent are i., ii. and iii. influenced by training?

Methodically, we utilize the original table of Polya (1945, xvi f.) that collects helpful questions to guide problem solvers through the four phases, It is reproduced here with questions numbered for reference and two insertions from Kilpatrick (1967, 45) in brackets shown in Table 1.

The following synopsis of this table with the heuristic dictionary of Polya (1945) shows that heuristic strategies occur mostly in the first two phases and that with the exception of *Working Forward* (which may be left out as a naturally occurring strategy) and *Transformation* all major strategies mentioned by Polya are comprised in the scheme. The order in which they appear in the table seems to be well-considered. (The other questions of the scheme address metacognitive issues.) (See Table 2)

Observe how many questions try to take previous knowledge into account – but empirical evidence suggests that more often than not problem solvers fail to do this properly. This is the rationale for our question 3.

The size and complexity of Polya's scheme also makes it pretty clear that one has to devise special means for heuristic training–it is not simply a matter of handing the list over to the students and let them apply it by trial and error (compare also Schoenfeld, 1985). Also, one would expect that acquisition of heuristic literacy takes place neither quickly nor easily–and even for experienced problem solvers it is certainly not straightforward to make good use of Polya's scheme, as subsequent

Table 1. Original table of Polya (1945)

UNDERSTANDING THE PROBLEM
First. You have to *understand* the problem. 1. What is the unknown? 2. What are the data? 3. What is the condition? 4. Is it possible to satisfy the condition? 5. Is the condition sufficient to determine the unknown? Or is it insufficient? 6. Or redundant? Or contradictory? 7. Draw a figure. 8. Introduce suitable notation. 9. Separate the various parts of the condition. Can you write them down?
DEVISING A PLAN
Second. Find the connection between the data and unknown. You may be obliged to consider auxiliary problems if an immediate connection cannot be found. You should obtain eventually a *plan* of the solution. 10. Have you seen it before? 11. Or have you seen the same problem in a slightly different form? *12. Do you know a related problem?* 13. Do you know a theorem that could be useful? *14. Look at the unknown!* And try to think of a familiar problem having the same or a similar unknown. *Here is a problem related to yours and solved before. Could you use it?* [15. Does it have a similar unknown?] 16. Can you use its result? 17. Could you use its method? 18. Should you introduce some auxiliary element in order to make its use possible? 19. Could you restate the problem? 20. Could you restate it still differently? Go back to definitions. If you cannot solve the proposed problem try to solve first some related problem. 21. Could you imagine a more accessible related problem? 22. A more general problem? 23. A more special problem? 24. An analogous problem? 25. Could you solve a part of the problem? 26. Keep only a part of the condition, drop the other part; how far is the unknown then determined, how can it vary? 27. Could you derive something useful from the data? 28. Could you think of other data appropriate to determine the unknown? [29. Could you derive something from the unknown?] Could you change 30. the unknown or 31. data or both if necessary, so that the new unknown and the new data are nearer to each other? 19. Could you restate the problem? 20. Could you restate it still differently? Go back to definitions. If you cannot solve the proposed problem try to solve first some related problem. 21. Could you imagine a more accessible related problem? 22. A more general problem? 23. A more special problem? 24. An analogous problem? 25. Could you solve a part of the problem? 26. Keep only a part of the condition, drop the other part; how far is the unknown then determined, how can it vary? 27. Could you derive something useful from the data? 28. Could you think of other data appropriate to determine the unknown? [29. Could you derive something from the unknown?] Could you change 30. the unknown or 31. data or both if necessary, so that the new unknown and the new data are nearer to each other? 32. Did you use all the data? 33. Did you use the whole condition? 34. Have you taken into account all essential notions involved in the problem?
CARRYING OUT THE PLAN
Third. *Carry out* your plan. 35. Carrying out your plan of the solution, *check each step*. 36. Can you see clearly that the step is correct? 37. Can you prove that it is correct? Looking Back
Fourth. *Examine* the solution obtained. 38. Can you *check the result?* 39. Can you check the argument? 40. Can you derive the solution differently? 41. Can you see it at a glance? 42. Can you use the result, or the method, for some other problem?

examples will illustrate. All in all, to assimilate Polya's scheme may well be viewed as an instance of instrumental genesis in the sense explained next. This may account for the disappointing results of Kilpatrick (1967).

Table 2. Synopsis with the heuristic diction of Polya (1945)

4. Induction
8. Draw a Figure
9., 25., 26. Decomposing and Recombining
18. Auxiliary Elements
14.- 18., 21.-24. Related Solved Problem, Auxiliary Problem
22. Generalization
23. Specialization
24. Analogy
26., (22., 23.) Variation
[27. Working Forward]
28., 29. Working Backward

In our ongoing research, we take these considerations into account as follows:

1. We plan to survey the heuristic development of two classes over a period of two years.
2. We will further the acquisition of heuristic skills by a suitable DGS usage.

In the main part of this paper, we will report on

1. How to design teaching units to foster heuristic literacy–the idea of heuristic reconstruction.
2. How to integrate DGS in the acquisition of heuristic skills–the concept of heuristic instrumentation.

We suppose that this rationale behind our approach can be fruitful also for others in designing DGS-based learning environment. Details on our research design can be found in Brockmann-Behnsen (2011).

THE CONCEPT OF HEURISTIC RECONSTRUCTION

Heuristic Reconstruction Means: Reconsider a problem and its solution so that one can pose it (and accompany the solution process if necessary) in such a way that enables students to make heuristic decisions on their own as often as possible. It is a design principle for teaching units that comprises of four guidelines bear a similarity to the four phases of Polya's scheme.

Guidelines for Heuristic Reconstruction

1. Start from a problem, often given in closed form (e.g. from a textbook), perhaps with a solution.
2. To prepare for educational use, pose the following *key questions*:
   a. What (intermediate) goals shall students reach within the course of processing the problem?
   b. What heuristic elements may help them to make process at a certain points? (Use e.g. Polya's list).
   c. Which of them can they find on their own? For which do they need a hint? Is the hint to be placed in the problem statement or better given when a certain state of problem processing is reached?
   d. What resources do they need to that end? Which of them can be taken as granted, which have to be supplied extra?
3. Use the answer to the key questions to rework and ameliorate the original task.
4. Observe how students interact with the heuristically reconstructed task and draw conclusions for the optimization of your task format.

Just as one may assume that good problem solvers do pass through Polya's phases, the designers of successful teaching units and textbooks will certainly have obeyed such guidelines anyway. Indeed such guidelines were already heeded by teaching practitioners and textbook authors with Polya's scheme in mind (Elschenbroich, personal communication). But since the majority of current teaching material unfolds only a small portion of the heuristic potential enclosed in the respective tasks it may be worthwhile to state these guidelines explicitly.

Heuristic reconstruction is kindred to the genetic approach: mathematics teaching "ought … to lead pupils slowly along the same path to higher ideas and finally to abstract formulations, as those that the human race in general followed from a naïve primitive state to knowledge at higher, abstract level." Klein (1924) The connection to heuristic becomes apparent in Wittmann's account: "In order to achieve mathematics teaching in which structural concepts, real mathematical problems and heuristic strategies are all given their full weight—the author would like to call the basic didactic conception here "genetic" —the choice of problems is of vital importance. Central problems will provide a starting point from which further problems and problem complexes can be generated. By these means, mathematical schemata and heuristic strategies can be learned and practiced." (Wittmann 1975, 194) The idea of empowering students to make their own decisions in the learning process is also implicit in Freudenthal's approach: "Good geometry instruction… means leading pupils to understand why some organization, some concept, some definition is better than another. Traditional instruction is different… All concepts, definitions and deductions are preconceived by the teacher" (Freudenthal, 1973).

Heuristic Reconstruction of a Concrete Example

Original Task: Can any triangle be divided into two isosceles triangles?

Arzarello et al. (2002) states it in the following way:

You are given a triangle ABC. Consider a point P on AB and the two triangles APC and PCB. Make a hypothesis about the properties of ABC which are necessary so that both APC and PCB are isosceles (such triangles are called "separable").

Compare this reconstruction of the original task to the one from Hölzl (1996) shown in Figure 1.

What changes occur? Both authors add several heuristic elements. Firstly, they both introduce suitable notation–one of Polya's basic heuristic advices. Then, while Arzarello's students have to produce a suitable drawing for themselves, Hölzl also provides two figures. Thereby he also unveils that the task is solvable at all–students may take this for granted, but consider Polya's question 4. Hölzl seems to decide that students are unlikely to pose it and provides the answer. Arzarello furthermore gives the hint that only some triangles are separable, by restating the task as: Find a necessary property for separable triangles. He may have reckoned that students' analysis of the original problem will not result in such a restatement. Furthermore his version only asks for a necessary condition–the original question affords to settle whether it is also sufficient. Arzarello may have deemed this too difficult for the students, though his instruction is more likely to lead to a sufficient condition by identifying a position of P where ABC becomes separable. Finally, Hölzl's two figures suggest that there are two cases of separable triangles–this could distract students from considering the possibility of other cases, but on the other hand may

Figure 1. Hölzl's triangle divison task

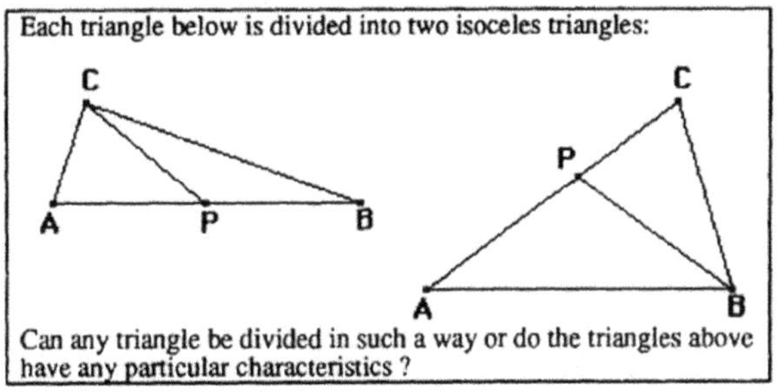

be helpful to draw their attention to the fact that there is not just one case, as in most other problems posed in school.

Altogether we recognize that both authors apply some sort of heuristic reconstruction, presumably according to their own unspoken guidelines. The result might be construed as mere simplification of the task–since the task has been proven empirically to be difficult, this adaption will certainly increase the possibility that students can make some progress on it. But from our point of view, such changes should not be executed a priori by the teacher, but explicitly and embedded in a learning situation, as to increase students' heuristic literacy.

Such a reconstruction affords to think though the original problem more thoroughly. Hefendehl-Hebeker & Hölzl (2000) provide a detailed analysis that addresses many aspects of our guidelines, including teaching experience. For space reasons we can give only a sketch of the heuristic reconstruction along the guidelines:

1. Students shall recognize that some, but not all triangles are separable. Then they shall come to discover that separable triangles fall into three (nondisjoint) classes and characterize these classes by suitable conditions (e.g. angle conditions).
2. Students shall produce examples and perhaps counter-examples for separability. They shall identify at least the class of right-angled triangles as separable.
   a. To come to necessary conditions, it is convenient to *Work Backwards*: Under what conditions can two isosceles triangles be *united* to a separable triangle? Then one can apply the heuristic *Systematically Distinguish Cases*: the triangles can be united along their bases or their sides respectively. From the four conceivable cases three actually do occur, as students can discover experimentally.
   b. To characterize the three cases by angle condition, only basic geometric knowledge is necessary. Salient, however, is some proof knowledge: E.g. the class of separable triangle that consist of isosceles triangles united along their respective sides contains just the right-angled triangles. One needs "if and only if" argumentations to show that each of the three combinatorial cases corresponds to an angle condition.
   c. To overcome possible difficulties, students can be supplied with examples as Hölzl does, or given a hint like Arzarello's that classes of separable triangles should be characterized by an equivalent condition. Probably it is necessary that the teacher provides the information that there are several cases and that they cannot be concisely characterized by *one* condition.
3. On the previous grounds we come to choose the following approach: First, the students shall explore the original situation freely for some time. Then a classroom discussion can address the following questions:

a. Can you find other separable triangles? (Expected answer: yes. Otherwise the teacher has to give some examples.)
b. Can you describe their properties in a way that makes it easy to decide whether a triangle belongs to this class of separable triangles? (Expected answer: If Hölzl's first case is found, it is recognized as a right triangle, but for the second case only few students will find a handy description.)
c. Does any triangle belong to these classes? (Expected answer: some students will have figured out a counter-example, where ones "sees" that it is not separable since it is not in the known classes of separable triangles (i.e. either the ones found by the students or Hölzl's two given classes), but without having thought of how to prove that there is no other way to divide it.)
d. Are their other separable classes? (Expected answer: the students will not have thought of this possibility.)
e. How can we finally make sure that we have found all separable cases? (This may be the moment, where some students could bring up the idea to revert the line of thought: "Let's assume, we have a separable triangle. What would the isosceles triangles into which it can be divided look like? What is the condition [Polya 3., 30., 31.!] that they could be united?" This paves the way to discover the combinatorial cases.)
f. Now that you think that you have found all possibilities to separate a triangle, can you describe precisely the resulting classes of triangles? (This induces the idea that the cases have different characterizations, which is unlikely to be taken into consideration by all but the most experienced students. Also, it suggests formulating an "if and only if"-statement for each case.)

This classroom dialogue may well be interrupted by extended periods of single or group work (e.g. to settle d.). Also, teaching material may be supplied if the process demands it (e.g. the teacher may provide worksheets that provide the formal proof structure necessary in f. for students that do not master this proof scheme).

This elaboration and instrumentation of the can easily be augmented to a whole teaching sequence, whose structure coincides quite naturally with Boero's phases:

1. Students are exposed to a problem situation. Investigating it–preferably in small groups-they produce some conjectures ("Not all triangles are separable-right triangles are.").
2. These conjectures should be exchanged and "mathematically streamlined"–at best via a classroom discourse. The teacher has to decide whether to disclose the existence of the third class–or even its characterization–to the students.

3. Depending on the strength of the group, one may leave it to students to explore the conjectures on their own and to collect possibly helpful arguments and tools for proving them–or do this also collectively.
4. Students can then set out in small groups to enchain arguments in order to validate or disprove their conjectures–and even transfer their lines of reasoning to the identify and handle the third class.
5. Again, it depends on the strength of the group, if one has the students organize their arguments–but since the problem is quite complex and demanding, one probably will do the "writing-up" together.
6. Then teacher may enhance the formal completeness of the proof by extra considerations (e.g.: "Are the three cases disjoint? What do we have to do to check this?").

This layout of the teaching unit shall give the students the freedom to make as many steps of the solution process as possible on their own. Anyway, it is fundamental that the important steps in this process are made conscious and emphasized as such by the teacher, be they found by the students or not. Some steps may be appreciated already during the solution process, but generally it seems worthwhile to look back after completion of the process and to highlight the function of the decisive steps. This is Polya's final phase – experience show that students seldom do it on their own (and it is also missing in Boero's model). But doing it collectively will in the course of time give them the possibility to acquire more and more heuristic strategies. We hypothesize that continuous efforts to make heuristics conscious will be the utmost effectiveness factor of heuristic reconstruction.

The Concept of heuristic instrumentation

What is the role of DGS in heuristic reconstruction? Its application may take place in Boero's phases III and IV, as will become clear by subsequent examples. But exactly how is DGS to be profitably instrumented in task? Arzarello et al. (2002) have investigated different functions of dragging and ordered them into a hierarchy of dragging modalities (Figure 2) in terms of "the twofold modalities, namely ascending and descending in which students interact with external representations".

Arzarello exemplifies the dragging modalities by his triangle separation task: Students may surmise that it is good idea to choose P as the midpoint of AB and construct a corresponding dynamic figure. Then they may drag C around freely to find cases of separable triangles – this is *wandering dragging*. If they try to drag C in order to realize examples of Hölzl's first case, they perform *dummy locus* or *lieu muet dragging* –the separable cases form a hidden locus (*lieu muet*), even without

Figure 2. Arzarello's hierarchy of dragging modalities

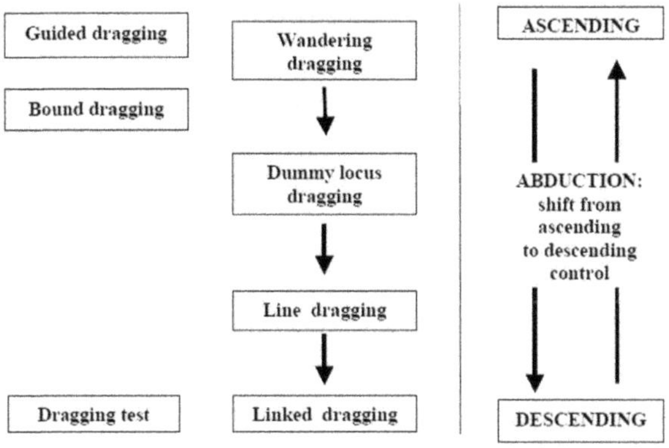

students being aware of it: Namely the triangle is separable when C is on the Thales circle of AB. If they mark such positions of C to make this locus visible, Arzarello speaks of *line dragging*. To validate the hypothesized connection, students may construct the circle, bind C to it and drag it around–*linked dragging*. (*Bound dragging* if the point was already constructed as semifree point on the circle.) Arzarello assumes that an essential factor in problem solving is the interplay of *ascending* and *descending control movements* between the visual and the theoretical level and suggests that these are facilitated by specific dragging modalities.

DGS may thus be seen as an apt means to foster heuristic strategies by suitable instrumentation of tasks. But one has to keep in mind that students will not necessarily utilize DGS on their own in the intended ways–Healy et al (1994) report "degoaling" phenomena: namely, the interaction with the software entails a shift of attention such that students tend to leave the original task out of consideration. They summarize their experience: "We knew from our Logo work that freedom to create and change one's own goal-degoaling-is double-edged. On the one hand, it allows pupils to appropriate the activities... On the other hand, it sometimes enables pupils to avoid interacting with mathematics at all!"

The observations of Hölzl (1996, 1999) point into the same direction: the interaction with the software often causes subtle differences between the tasks' intended goals and the situated knowledge acquired by the students.

Therefore, not only the utilization of Polya's scheme, but also of other instruments, particularly DGS needs thorough consideration. We define *heuristic instrumentation* as the design phase of a teaching unit, were such considerations take place–naturally on the ground of information about the learners' heuristic literacy as well as their DGS proficiency. The concept is illustrated best with an example.

Example for Combined Heuristic Reconstruction and Instrumentation

Original Task: (from Elschenbroich (1997, 63) Draw an equilateral triangle. Choose an arbitrary point inside the triangle and measure the distances x, y, z to the sides. What can you say about the sum x+y+z? Is there any relation to other data in the triangle?

This task was criticized by Hölzl (2001): "Such DGS activity can hardly be expected to yield more than empirical confirmation, nor is it likely to point to a promising solving strategy. In fact, such a strategy necessitates a mental restructuring of the triangle into appropriate equilateral triangles, thus requiring a purposeful view that already draws on heuristic skills but hardly comes to mind by just dragging P around." Hölzl also takes into account that "Elschenbroich mentions in a footnote that one could argue whether one had not better posed the problem in a more open manner, for example: "What can you say about x, y, z?" In doing so, however, the heuristic role of DGS would not be significantly enhanced: Dragging P while watching three simultaneously changing variables does not suggest considering their sum." One may readily recognize the considerations of Elschenbroich as well as of Hölzl as attempts to reconstruct the problem and improve its instrumentation to enlarge the heuristic value of the problem. From the point of view of our conceptual framework, we can propose the following means to that end: Accept that the shift of focus towards x+y+z is unlikely to be performed by the students on themselves and try to gain the heuristic value in another way:

Reconstructed Example: Let ABC be an equilateral triangle and let P be on or inside ABC. Let D, E and F be feet of the perpendiculars from P to AB, AC and BC and $S(P):=|DP|+|EP|+|FP|$. What can you say about S(P)?Consider special cases!

Indeed, this restatement of the task it suitable namely for phase I and for utilization of DGS: namely dragging P inside or on ABC, one will readily see that S(P) is constant, as prior. But to establish its precise value, specialization can now be utilized: namely, by dragging P to a vertex, say C, one discovers that in this case $S(P)=|CE|=:h_{ABC}$ equals the length of an altitude of ABC.

Reduction to special cases is also a successful strategy for phase III and IV. To that end, the helpful notation S(P) was introduced to the problem since it facilitates then reduction from more general P to more special P (See Figure 3). For instance, to reduce the case P on AB to the special case P=A, one may consider the triangle PBQ, where Q is the intersection point of BC with the parallel to AC through P. This observation can be completed to a valid proof in phase IV if one considers that for the corresponding sum in PBQ we have $S^*(P) = |PF| = h_{PBQ}$ and uses this to rewrite $S(P) = |PE| + |PF| = d(AC, PQ) + S^*(P) = h_{ABC}$.

Figure 3. Intermediate step in the stepwise reduction of the task to special cases

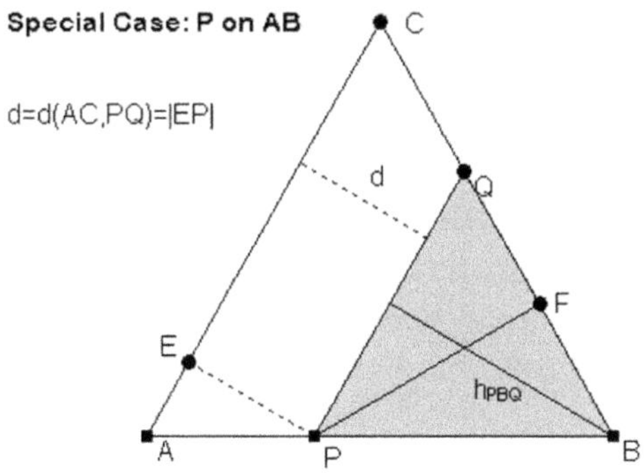

However this line of thought is still unlikely to be discovered by students on their own. Now DGS can be utilized as mediating tool by the teacher: He may advise the students to consider PBQ (or, less directively: an equilateral triangle with vertex P) and its connection to the given problem. By dragging P along AB, the students sees the case P=A as limit case and this may spark off the idea to reduce the case P on AB to the special case P=A Note however, that the tool is not directly helpful to find the prior representation of S(P) in terms of S*(P). This and especially the introduction of the new notation S*(P) has certainly to be introduced by the teacher in a suitable way and in the right moment. This shows clearly the limits of what can be achieved by pre-formulated instruction: a worksheet may provide a suitable starting point for a problem solving process, but the teacher has to monitor closely the subsequent group work and must be prepared to intervene in turning points when students are likely to get stuck.

Heuristic Instrumentation by Use of Electronic Worksheets

To support teachers in handling such difficulties, Elschenbroich (2001) introduced the concept of EiWos: Electronic interactive Worksheets. These provide the learners with pre-constructed figures to avoid the pitfalls of "geometrical programming" (delays by construction errors, necessity of DGS specific knowledge, "degoaling") and allow them to concentrate on problem solving.

In Elschenbroich's EiWos, tasks are usually broken up into a sequence of well-considered steps, with detailed dragging instructions if necessary–as in the reconstruction of the previous task (ibid., http://www.ielab.de/eiwos-3.htm). This will ensure that average teachers can guide average learning groups through the solution of a non-routine problem. The necessary heuristic strategies and their instrumentation, however, are all preconceived by the author–so it seems questionable whether they will be assimilated in such a way that they can be utilized in more open learning situation, where no way of task processing is prescribed. It deems us better to let the teacher support the choice of heuristics, but make the selection transparent to the learners.

We shall methodically investigate this issue by contrasting a learning group that sticks closely to Elschenbroich's EiWos with a parallel group that receives augmented instruction with respect to heuristic instrumentation. Namely, to tackle with research question IV., we currently survey two treatment groups (regular classes, at present in grade 8) for two years:

- **Explicit Training (ET):** This class will work with "opened" EiWos (i.e. they can utilize same heuristics like introducing auxiliary elements on their own).
- **Implicit Training (IT):** This class will use the original EiWos with prescribed heurisms.

Figure 4 shows the first three EiWos from Elschenbroich & Seebach (2002). Italicized instructions are omitted in ET, A1-03 c) is omitted in IT.

Figure 4. Three EiWos to discover and prove Thales' theorem

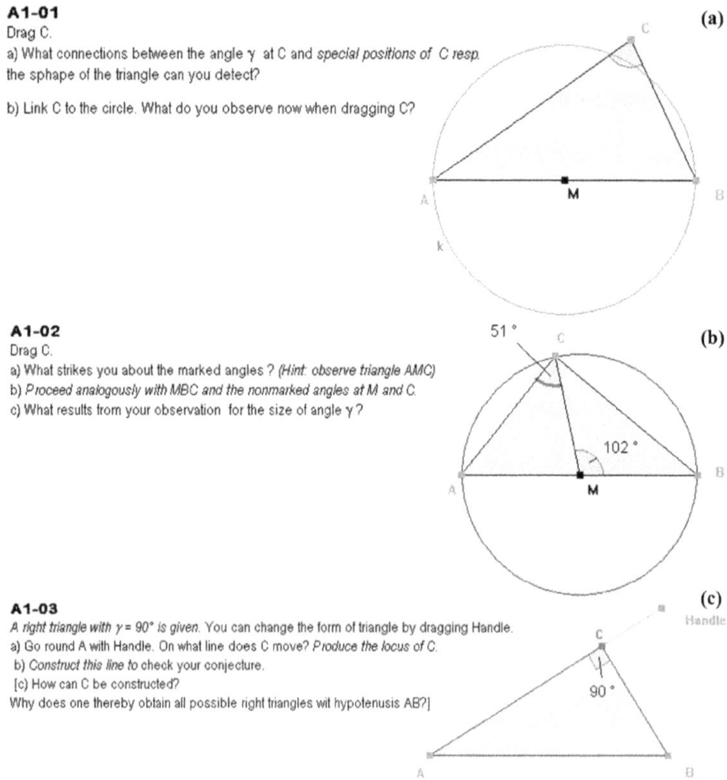

These EiWos can be viewed as a heuristic reconstruction of Thales' theorem. Our treatment starts with a teaching unit that utilizes them. It can be succinctly sketched using Boero's phases:

Lesson Plan for a Teaching Unit on Thales' Theorem

1. Free exploration (in partner work) of
   - **Task 1:** Consider a circle k with diameter AB and a free point C. Construct the triangle ABC. How does the angle at C vary with the position of C?

   A1-01 realizes precisely this task in a DGS environment!
2. Carve the theorem's formulation out of students' conjectures in classroom discourse:
   - **Theorem (Thales):** Given a circle k with diameter AB and an arbitrary point C, then the triangle ABC is right-angled at C if C lies on k.

3 & 4. Let students explore the contents of the theorem, find appropriate arguments and try to form a deductive chain. Some aid for these steps is provided by A1-02.

5. Streamline the enchained arguments to produce a valid proof in classroom discourse.
6. The teacher may throw up the question whether the deductive chain can be reverted – whereby one would formally obtain the converse of the theorem.

But in most cases he would rather decide to let students explore on their own whether the converse is also true, thereby returning to I. A1-03 supports this approach.

A Refined Classification of Dragging Modalities for Research and Teaching

As we advocated before making heurism usage conscious and deliberate, we would now like to propose the same for the acquisition of instrumented DGS techniques. But how to achieve this? The question whether dragging modalities should be taught has not yet been addressed in the literature. However, for best results it will be necessary in classroom discourse as well as in instructional texts (e.g. A1-01 b) and A1-03a)!) to linguistically distinguish between the dragging modalities in an efficient way–this is Polya's heuristic *Introduce suitable notation,* and Arzarello gives no empirical validation of his model and does not claim that it is all-encompassing. Especially "bound dragging" occurs in several manifestations that should be distinguished.

We therefore propose the following refinement taxonomy: *Modalities of dragging* are first construed as consisting of *dragging acts* that are performed in various *articulations.*

Concerning the mere dragging act, we propose a systematic classification along two dimensions:

1. One can distinguish, *what kind of point is dragged*: a base point, a semifree point (i.e. one that is bound to a line) or a dependent point (its position is determined by the given construction, so it cannot be moved at all).
2. One can distinguish *different ways of dragging*: Namely something can be dragged away from some object or attempted to be moved in a certain direction or to some position. Then, one might want to move an object along a geometric locus – be it visible or imagined. Finally, a move may not follow any apparent plan at all.

This systematics is exhaustive in the following sense:

1. The dragged point falls necessarily into the previous categories.
2. Any dragging will either *preserve* a certain geometric relation, *establish* a new one or *terminate* an existing one. Such a relation may equivalently be viewed as a geometric property or as a locus condition.

Thus we obtain the *taxonomy of dragging acts* shown in Table 3.

Some of Arzarello's modalities fit into this scheme. These are italicized - new dragging acts are also boldfaced.

Table 3. Taxonomy of dragging acts

Dragging acts	**How is it dragged ?**			
What is dragged?	**Relation-preserving**	**Relation-terminating**	**Relation-establishing**	**Without apparent aim**
Base point	Through a region or locus or to keep a geometric property (*lieu muet dragging*)	***Dragging away*** from some object or to abandon a geometric property	***Dragging toward*** some object or to realize a geometric property (*Guided dragging*)	*Wandering dragging*
Semifree point	Ditto	Ditto	Ditto	*Bound dragging*
Dependent point	***Powerless dragging*** to check whether a point can be moved in these ways or to demonstrate that this is impossible			

Sometimes terminating an existing relation and establishing a new one cannot be clearly distinguished or have to occur simultaneously (see example). Thus it makes sense to combine them to *relation-changing dragging.*

Now each of these dragging acts may occur in various *articulations,* namely:

- *What is observed* during dragging? Is it the dragged point itself or is it a point or a configuration that depends on it?
- Is the observed object *merely dragged* or is its dragging behavior made visible in some way:
  - By *marking s*elected positions with newly constructed points.
  - By *tracing* the dragging path of the observed object.
  - By *producing* its locus.
- In what phase does the dragging act occur? We can distinguish:
  - Explorative use in an ascending phase.
  - Verifying use in a descending phase.

The salience of the refined scheme will become evident by studying the instrumentalisation of DGS in the previous task 1: In exploring the situation, the problem solver would presumably start by just moving around C in some way or other (i.e. wandering dragging (1)). After some time (or by a hint) it may occur to him that inside k, the angle is always acute. To check this conjecture, one would move C consciously inside k and watch the angle–Arzarello's dragging test, here construed as relation-preserving dragging act in verifying use (2). To see what happens outside C, one would move C across k–relation-changing dragging in explorative use. (3). Moving it around there exploratively (4) would be Arzarello's guided dragging–but Arzarello would probably dub it a dragging test if it is used to confirm the observed obtuseness. By binding C to k, the limit case can be established–then bound dragging in either explorative or verifying use occurs (5). The effect of the binding would be tested or demonstrated by powerless dragging (Figure 5).

Further modalities arise if the same dragging acts in another articulation–this will occur if one reconstructs the problem slightly by removing k. Then one might ask: When is the angle at C acute, when is it obtuse? This may again be explored via wandering dragging. But then the problem solver becomes perhaps interested in figuring out when the angle is exactly a right angle. To that end he will use guided dragging. In order to explore the locus of all these "boundary points" he might try to move C so that the angle remains constant (i.e. lieu muet dragging). It may be helpful to mark points on this locus to make it visible. Utilizing the trace like in the figure would be another possible articulation of the same dragging act.

Figure 5. A dragging path whilst exploring the Thales configuration

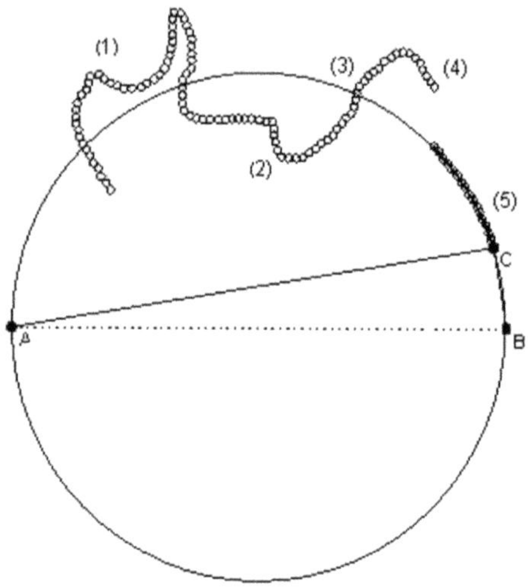

So far, the dragged point was also the object, whose movement was to be observed. By relaxing this condition, one arrives at some additional dragging modalities. These would readily apply in the following

- **Task 2:** Given a triangle ABC, when does its orthocenter H lie inside ABC, if you vary the position of C?

Observe that these examples only exhibit heuristic use of DGS in Boero's phases I and II. A heuristic usage in phase III occurs in the following

- **Task 3:** If the problem solver of Example 1 is to formulate and prove a theorem on his own, he may arrive at Thales' theorem. By using Polya's heuristic *Auxiliary Elements* he may introduce the center M of k and draw MC that dissects the angle at C. Using wandering dragging, he can observe that its parts are always equal to the angles at A and B (*Decomposing and Recombining* or *Invariance Principle*). Using the *Related Solved Problem* that the measures of the interior angles of a Euclidean triangle always add up to 180 degrees the proof can be finished. Some of these steps are supported in A1-02.

The Complexity of Heuristics

Our schemes of heuristic reconstruction and instrumentation serve several purposes–one is to secure that in the analysis of tasks no possibilities to reconstruct and instrument are overlooked. This will be helpful with demanding tasks. Also, our schemes help to clarify what makes heuristic work difficult. For instance it can be readily seen that applying Polya's scheme is not a straightforward affair by trying to reconstruct the following example for *Working Backwards*:

Primary Task: (Hölzl, 1996) Consider a triangle ABC inscribed in a circle k with A varying on k Show that the locus M of the midpoint M of AB is a circle (Figure 6).

Its reconstruction may again start with discarding the hypothesis to give students the opportunity to go through Boero's phases I and II. The task is then suitable to illustrate the use of several of Polya's original questions-phase III can even be framed as an inner dialogue in the spirit of Polya's teacher-student-dialogues:

What is the unknown! A circle. *What is the condition?* Every circle is determined by center and radius. *Try to think of a familiar problem having the same or a similar unknown.* Many locus problems have a circle as their solution. *Could you imagine a more accessible related problem?* We look for a circle's center. The given circle has also a center. *Introduce suitable notation.* It is the circumcenter O of ABC. *Could you restate the related problem?* If ABC is inscribed in a circle k, the centre of k coincides with the circumcenter O. *Could you change the unknown so that the new unknown and the new data are nearer to each other?* To that end, one may consider that OMB is also a triangle that is inscribed in a circle. *Could you restate the problem?* One could ask: how is the unknown circle determined by OMB? *Could you restate it still differently?* Of course, the unknown circle is the circumcircle of OMB. But that does not seem to be directly helpful. *Could you imagine a more accessible related problem? … Did you use all the data?* Indeed, OMB is inscribed in a special way: OB is a diameter of the unknown circle. This means that one can apply Thales' theorem!

Figure 6. Hölzl's locus task may be explored dynamically

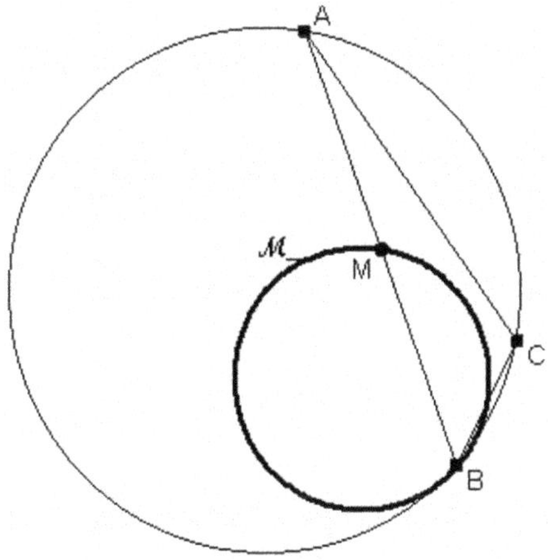

Now the approach to *Work Backwards* starts to look promising: Using the converse of Thales' theorem, one can prove the desired conclusion if one is able to prove that angle OMB is always a right angle. This goal can be achieved in the same vein: using the hypothesis as premise, one derives that OMB is the mirror image of OMB with respect to the reflection axis OM. Thus AOB is isosceles. By inverting this line of thought one readily arrives at the desired hypothesis, thereby completing phase IV successfully (See Figure 7).

However, one sees that the solution path jumps backwards and forward all over Polya's scheme. It is unlikely that actual learning trajectory proceed in this way–but it will be fruitful to compare them after the solution has been found. Looking back in this way may serve to illustrate the value of heuristic and contribute to remember some of the questions more easily in future problems (Figure 8).

How can DGS be utilized in solving this problem? If one sticks to the solution path prior, it can only play a minor role (e.g. step 21. is supported if one observes the invariance of the right angle at M by exploratively dragging C along k). This occurs indeed, see Hölzl (1996). But furthermore the dynamic visualizations entail a conceptual reframing of the problem situation: the solvers view it dynamically instead of statically–namely it occurs to them that M arises from A by a dilatation with center B. Now it becomes pretty clear that the locus must be a circle–dilatations preserve distances!

This empirical observation illustrates how far-reaching the influence of DGS on a proof process can be.

Figure 7. Heuristic augmentation of Högl's configuration may instigate the proof process

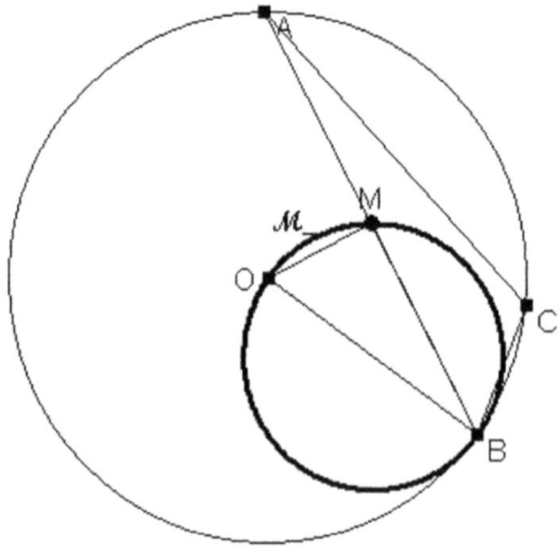

Figure 8. Solution path for Hölzl's task in terms of Polya's scheme

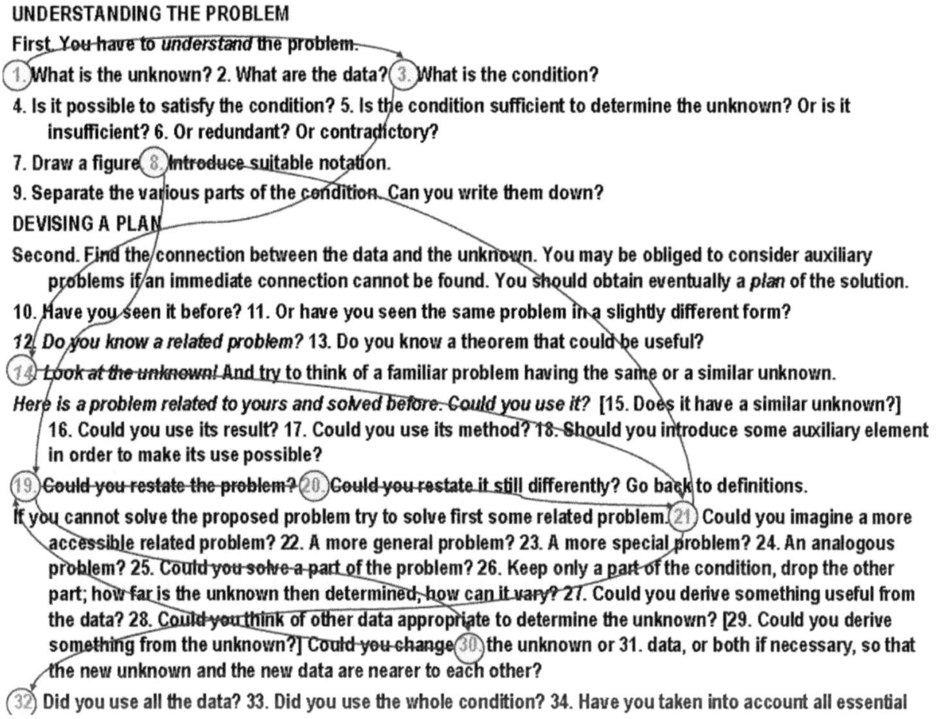
UNDERSTANDING THE PROBLEM

First. You have to *understand* the problem.

1. What is the unknown? 2. What are the data? 3. What is the condition?

4. Is it possible to satisfy the condition? 5. Is the condition sufficient to determine the unknown? Or is it insufficient? 6. Or redundant? Or contradictory?

7. Draw a figure. 8. Introduce suitable notation.

9. Separate the various parts of the condition. Can you write them down?

DEVISING A PLAN

Second. Find the connection between the data and the unknown. You may be obliged to consider auxiliary problems if an immediate connection cannot be found. You should obtain eventually a *plan* of the solution.

10. Have you seen it before? 11. Or have you seen the same problem in a slightly different form?

12. *Do you know a related problem?* 13. Do you know a theorem that could be useful?

14. *Look at the unknown!* And try to think of a familiar problem having the same or a similar unknown.

Here is a problem related to yours and solved before. Could you use it? [15. Does it have a similar unknown?] 16. Could you use its result? 17. Could you use its method? 18. Should you introduce some auxiliary element in order to make its use possible?

19. Could you restate the problem? 20. Could you restate it still differently? Go back to definitions.

If you cannot solve the proposed problem try to solve first some related problem. 21. Could you imagine a more accessible related problem? 22. A more general problem? 23. A more special problem? 24. An analogous problem? 25. Could you solve a part of the problem? 26. Keep only a part of the condition, drop the other part; how far is the unknown then determined, how can it vary? 27. Could you derive something useful from the data? 28. Could you think of other data appropriate to determine the unknown? [29. Could you derive something from the unknown?] Could you change 30. the unknown or 31. data, or both if necessary, so that the new unknown and the new data are nearer to each other?

32. Did you use all the data? 33. Did you use the whole condition? 34. Have you taken into account all essential

AN ADVANCED CASE STUDY

Finally, we want to elucidate the virtues of an elaborate dragging practice in an advanced problem solving situation. Some years ago we were able to observe the work of Deinhardt, a heuristically talented primary teacher student. For her final examination thesis we instructed her to delve into an example already reconstructed by Hölzl:

Primary Task: *Show that the locus $\mathscr{H}$ of the orthocentre H of a triangle ABC inscribed in a circle k with C varying on k is the reflection of k by BC.*

Traditionally, one proves this via transformation geometry, but runs into trouble when in phase IV trying to justify rigorously the visually obvious existence of the reflection. Since the needed background is falling off in schools anyway, this should not be reckoned as a well suited entrée problem-but even worse: Hölzl (2001) reports empirical evidence that for many students even the hypothesis itself does not seem to be worth noting: since C varies on a circle, they deemed it natural that the locus is

Figure 9. Four instances of the locus of the orthocentre when B moves away from k

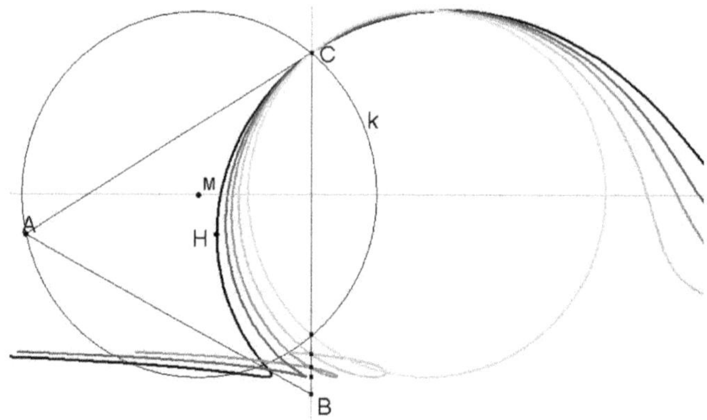

also a circle. In order to elucidate its peculiarity, Hölzl has proposed to untie B from k and then explore the situation–thereby prescribing the strategy of generalization! If one drags B in this way, the circle gives way to an unfolding variety of curves – Figure 9 shows four of them emerging from the circle when moving B away from k.

This astonishing behavior may stimulate analytical work to explain the peculiarity of the special case. Specifically, Deinhardt was instructed to investigate:

1. How can the manifold of occurring curves be ordered and described?
2. Can something interesting be said about this manifold?

Figure 10. Deinhardt's construction of a universal family of orthocentre loci

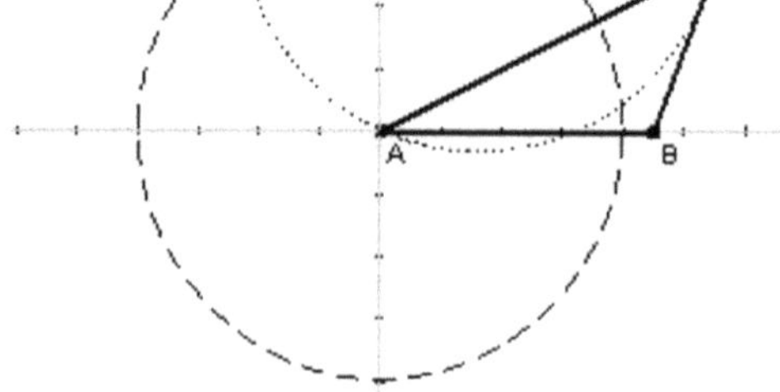

The task was set quite broadly to allow for heuristic work. Deinhardt solved it so splendidly that the results were published (Deinhardt 2000)! We can only sketch them:

1. Deinhardt succeeds to construct a universal family $\mathscr{H}_{M,B}$ of orthocentre loci that realizes all possible configurations of A,B, C and k up to similarity (Figure 10). With coordinates it can be described as follows: A is fixed to the origin, B is linked to the positive x-axis. C then varies on a circle k of radius 1 whose centre M itself may also be varied on the unit circle. Thus one successively covers all possible configurations by varying M and B. Deinhardt next orders them according to their *gestalt:* do they have asymptotes, singularities....? For brevity of space, we must refrain from explaining her taxonomy.

Introducing coordinates then allows Deinhardt to utilize the heuristic strategy *Transformation*: Via the Cartesian correspondence the geometric construction of $\mathscr{H}$ can be translated to an algebraic description: Elementary analytic geometry yields a parameterization of $\mathscr{H}$. By eliminating the parameter one gets an equation for $\mathscr{H}$ that depends on M and B.

2. Deinhardt chooses to inspect the occurring singularities more closely. The exploration of $\mathscr{H}_{M,B}$ leads her to the observation that $\mathscr{H}$ can have at most one singular point: either a node or a cusp. As can been seen previously, nodes occur quite frequently, whereas cusps are more rare. When do they occur?

Figure 11. Beginning the exploration of the cusp locus

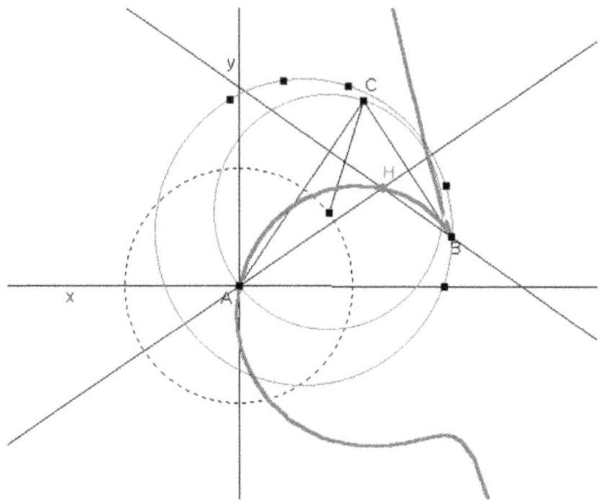

Figure 12. Discovery of the cusp of the cusp locus

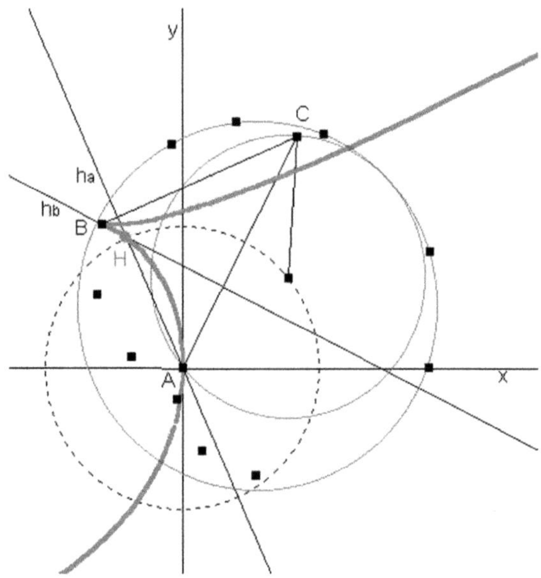

Figure 13. The cusp locus turns out to be a cardioid

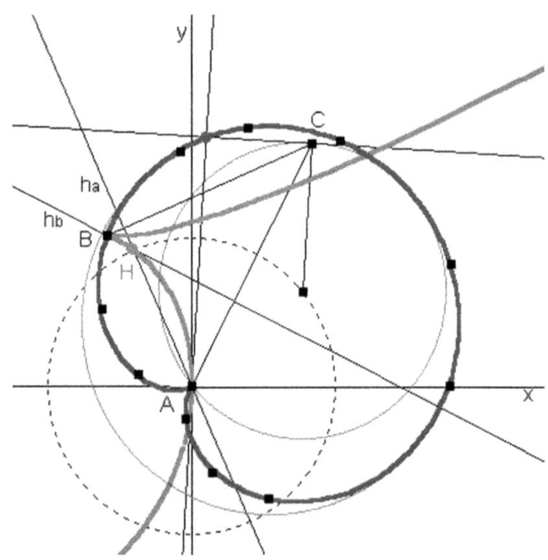

For a concise description of the cusp case, Deinhardt unties B from the x-axis. By lieu muet dragging, she finds some positions of B such that $\mathscr{H}_{M,B}$ has a cusp. It looks as if these points lie on a circle (Figure 11).

But further investigation disproves this conjecture: Near A the locus of cusp points itself seemingly has a cusp! (Figure 12)

By meticulously exploring the situation geometrically and analyzing the results algebraically, Deinhardt succeeds in uncovering a surprising result: It turns out that the locus of cusp points B (i.e. position of B such that $\mathscr{H}_{M,B}$ has a cusp) is a cardioid! (Figure 13)

Deinhardt is able to again utilize *Transformation* to confirm here findings by analytic geometry. She thus gives a stunning example for a double "instrumental genesis": By applying advanced heuristic strategies and instrumenting them with refined DGS utilizations schemes (lieu muet dragging, binding and untying points), Deinhardt is able to achieve far more than one would have ever reckoned.

DGS can here be seen to advance the solution process by alleviating the switch between *ascending* and *descending control movements* sensu Arzarello as well as the change of enactive, iconic and symbolic *representation modes* sensu Bruner–instead of a detailed analysis we must restrain ourselves to a schematic diagram (Figure 14).

CONCLUSION

The example of Deinhardt shows how heuristic problem solving is carried forward by skillful utilization of various dragging modalities. It is not meant to suggest that most or many students will reach this level of mastery by our heuristic training. But we will investigate the long-term impact of integrating EiWos into daily classroom

Figure 14. Control movements and mode changes during the utilization of DGS in problem solving

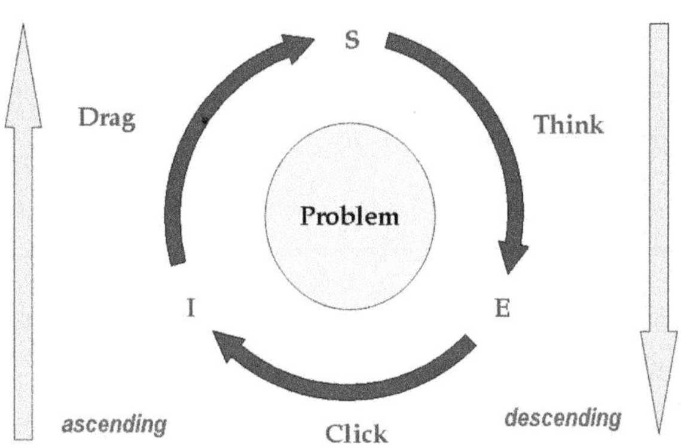

practice, using standard as well as non-standard problems as benchmarks. And we do expect that this training will significantly increase the heuristic literacy of our students.

REFERENCES

Artigue, M. (2002). Learning mathematics in a CAS environment: The genesis of a reflection about instrumentation and the dialectics between technical and conceptual work. *International Journal of Computers for Mathematical Learning*, *7*, 245–274. doi:10.1023/A:1022103903080.

Arzarello, F., Micheletti, C., Olivero, F., Robutti, O., Paola, D., & Gallino, G. (1998). A model for analysing the transition to formal proofs in geometry. In *Proceedings of PME 22*. Stellenbosch, South Africa: PME Press.

Beckmann, A. (2001). Probleme beim beweisenlernen–DGS als lösung? In Elschenbroich, Gawlick, & Henn (eds.), Zeichnung–Figur–Zugfigur. Hildesheim, Germany: Franzbecker.

Boero, P. (1999). Argumentation and mathematical proof: A complex, productive, unavoidable relationship in mathematics and mathematics education. *International Newsletter on the Teaching and Learning of Mathematical Proof, 7/8*.

Brockmann-Behnsen, D. (2011). A long-term educational treatment using dynamic geometry software. In *Proceedings of ICTMT 10*. Portsmouth, England: Open University Press.

Deinhardt, A. (2000). Die ortskurven des höhenschnittpunktes. *Vechtaer Fachdidaktische Forschungen und Berichte, 3*.

Elschenbroich, H.-J. (1997). Tod des beweisens oder wiederauferstehung? In Hischer, H. (Ed.), *Computer and Geometrie. Neue Chancen für den Geometrieunterricht?* Hildesheim, Germany: Franzbecker.

Elschenbroich, H.-J. (2001). *Electronic interactive worksheets–A visual and dynamic way of learning geometry*. Montreal: Cabri World.

Elschenbroich, H.-J., & Seebach, G. (2002). *Dynamisch geometrie entdecken. Klasse 7, 8, 9, 10*. Rosenheim, Germany: CoTec.

Forehand, G. (1967). *Individual differences in problem solving processes of college students*. Pittsburgh, PA: Carnegie Mellon University.

Freudenthal, H. (1973). *Mathematics as an educational task.* Dordrecht, Netherlands: Reidel.

Gawlick, T. (2002). On dynamic geometry software in the regular classroom. *Zentralblatt für didaktik der mathematic, 34*(3), 85-92.

Gawlick, T. (2003). Utilizing DGS as heuristic tool in introductory calculus courses. In *Proceedings of ICTME 2*. Beirut, Lebanon: ICTME Press.

Gawlick, T. (2005). Connecting arguments to actions–Dynamic geometry as means for the attainment of higher van hiele levels. *ZDM. Zentralblatt für Didaktik der Mathematik, 37*(5).

Hayes, J. (1966). Memory, goals, and problem solving. In Kleinmuntz, B. (Ed.), *Problem Solving: Research, Method, and Theory*. New York: Wiley.

Healy, L., Holzl, R., Hoyles, C., & Noss, R. (1994). Messing up. *Micromath, 10*(1), 14–16.

Hefendehl-Hebeker, L., & Hölzl, R. (2000). Zerlegbare Dreiecke im problemorientierten Unterricht. In Baptist (ed.), Mathematikunterricht im Wandel. Bamberg, Germany: Buchner.

Hölzl, R. (1996). How does 'dragging' affect the learning of geometry? *International Journal of Computers for Mathematical Learning, 1*(2), 169–187. doi:10.1007/BF00571077.

Hölzl, R. (1999). *Qualitative unterrichtsstudien zur verwendung dynamischer geometrie-software*. Augsburg, Germany: Wissner.

Hölzl, R. (2001). Using DGS to add contrast to geometric situations. *International Journal of Computers for Mathematical Learning, 6*(1), 63–86. doi:10.1023/A:1011464425023.

Kilpatrick, J. (1967). *Analyzing the Solution of Word Problems in Mathematics: An Exploratory Study.* (Dissertation). Stanford, CA, Stanford University.

Klein, F. (1924). *Elementarmathematik vom höheren Standpunkte aus*. Berlin: Springer. doi:10.1007/978-3-642-90853-8.

Laborde, C., & Laborde, J.-M. (1992). Problem solving in geometry: From microworlds to intelligent computer environments. In Ponte (ed.), Mathematical Problem Solving and New Information Technology. Berlin: Springer

Laborde, C., & Laborde, J.-M. (1995). What about a learning environment where euclidean concepts are manipulated with a mouse? In diSessa, A. (Ed.), *Computers for Exploratory Learning*. Berlin: Springer.

Newell, A., & Simon, H. (1972). *Human problem solving*. Englewood Cliffs, NJ: Prentice-Hall.

Olivero, F. (1999). Cabri-Géomètre as a mediator in the process of transition to proofs in open geometric situations. In: W. Maull & J. Sharp (Eds.), *Proceedings of ICTMT 4*. Plymouth, UK: University of Plymouth.

Pólya, G. (1945). *How to solve it*. Princeton, NJ: Princeton University Press.

Rott, B. (2011). Problem solving processes of fifth graders: An analysis. In *Proceedings of PME 35*. Ankara, Turkey: PME.

Rott, B., & Gawlick, T. (Manuscript submitted for publication). Sind mathematische Heurismen trainierbar? *Unterrichtswissenschaft*.

Schoenfeld, A. (1985). *Mathematical problem solving*. Orlando, FL: Academic Press.

Schoenfeld, A. (1992). Learning to think mathematically: Problem solving, metacognition, and sensemaking in mathematics. In Grouws, D. (Ed.), *Handbook for Research on Mathematics Teaching and Learning*. New York: MacMillan.

Van Hiele, P. M. (1986). *Structure and Insight*. Orlando, FL: Academic Press.

Vérillon, P., & Rabardel, P. (1995). Cognition and artifacts: A contribution to the study of thought in relation to instrumented activity. *European Journal of Psychology of Education*, *10*(1), 77–101. doi:10.1007/BF03172796.

Wittmann, E. (1975). Matrix strategies in heuristics. *International Journal of Mathematical Education in Science and Technology*, *6*(2), 187–198. doi:10.1080/0020739750060207.

ADDITIONAL REFERENCES

Baccaglini-Frank, A. (2011). Abduction in generating conjectures in dynamic geometry through maintaining dragging. In Swoboda (ed.), *Proceedings of CERME 7*. Rzeszow, Poland: CERME Press.

Balacheff, N. (1987). Processus de preuve et situations de validation. *Educational Studies in Mathematics*, *18*(2), 147–176. doi:10.1007/BF00314724.

Clements, D., & Battista, M. (1994). Computer environments for learning geometry. *Journal of Educational Computing Research, 10*(2), 173–197. doi:10.2190/8074-298A-KTL2-UQVW.

De Villiers, M. (1997). The role of proof in investigative, computer-based geometry: Some personal reflections. In Schattschneider & King (eds.), Geometry Turned On! Washington, DC: MAA.

De Villiers, M. (2003). *Rethinking proof with sketchpad 4*. Emeryville, CA: Key Curriculum Press.

De Villiers, M. (2007). Some pitfalls of dynamic geometry software. *Teaching & Learning Mathematics, 4*, 46–52.

Defouad, B. (2000). Etude de Genèses Instrumentales Liées à l'Utilisation d'une Calculatrice Symbolique en Classe de Première S. (Doctoral thesis). Paris, France, University of Paris.

Drijvers, P. (2000). Students encountering obstacles using CAS. *International Journal of Computers for Mathematical Learning, 5*(3), 189–209. doi:10.1023/A:1009825629417.

Guin, D., & Trouche, L. (1999). The complex process of converting tools into mathematical instruments: The case of calculators. *International Journal of Computers for Mathematical Learning, 3*(3), 195–227. doi:10.1023/A:1009892720043.

Hölzl, R. (1994). *Im Zugmodus der Cabri-Geometrie. Interaktionsstudien und Analysen zum Mathematiklernen mit dem Computer*. Weinheim, Germany: Deutscher Studien Verlag.

Laborde, C. (1998). Visual phenomena in the teaching/learning of geometry in a computer-based environment. In Mammana, C., & Villani, V. (Eds.), *Perspectives on the Teaching of Geometry for the 21st Century*. Dordrecht, Netherlands: Kluwer.

Laborde, C. (2000). Dynamic geometry environments as a source of rich learning contexts for the complex activity of proving. *Educational Studies in Mathematics, 44*(1-3), 151–161. doi:10.1023/A:1012793121648.

Laborde, C. (2001). Integration of technology in the design of geometry tasks with cabri-geometry. *International Journal of Computers for Mathematical Learning, 6*(3), 283–317. doi:10.1023/A:1013309728825.

Laborde, C., Kynigos, C., Hollebrands, K., & Sträßer, R. (2006). Teaching and learning geometry with technology. In Gutierrez, A., & Boero, P. (Eds.), *Handbook of Research on the Psychology of Mathematics Education*. Rotterdam, Netherlands: Sense Publishers.

Noss, R., & Hoyles, C. (1996). *Windows on mathematical meanings*. Dordrecht, Netherlands: Kluwer. doi:10.1007/978-94-009-1696-8.

Olivero, F. (2002). *Proving within Dynamic Geometry Environments*. (Ph.D. Thesis). Bristol, UK, Graduate School of Education, Bristol.

Polya, G. (1954). *Mathematics and plausible reasoning. Induction and analogy in mathematics, 1*. Princeton, NJ: Princeton University Press.

Schumann, H. (1991). Interactive generalizing of geometric configurations. *International Journal of Computers for Mathematical Learning*, *22*(6), 953–963.

Sträßer, R. (2001). Chancen und Probleme des Zugmodus. In:Elschenbroich, Gawlick, & Henn (eds.), Zeichnung–Figur–Zugfigur. Hildesheim, Germany: Franzbecker.

Sweller, J. (1990). On the limited evidence for the effectiveness of teaching general problem solving strategies. *Journal for Research in Mathematics Education*, *21*(5), 411–415. doi:10.2307/749398.

KEY TERMS AND DEFINITIONS

Dragging Modalities: Dragging acts that are performed in various articulations.

EiWos: Electronic interactive Worksheets backing the heuristic instrumentation of problems.

Heuristic Instrumentation: Method to integrate DGS in the acquisition of heuristic skills by conscious use of dragging modalities.

Heuristic Literacy: Students ability to apply heuristics.

Heuristic Reconstruction: Method to design teaching units to foster heuristic literacy utilizing Polya's problem solving scheme.

Section 2
The Practice

Chapter 4
Dynamical Mathematical Software:
Tools for Learning and for Research

Samer Habre
Lebanese American University, Lebanon

ABSTRACT

Understanding mathematical concepts is many-folded. Traditional mathematics mostly emphasizes the algebraic/analytical aspect of a problem with minimal reference to its graphical aspect and/or numerical one. In a modern learning environment, however, multiple representations of concepts are proving to be essential for the teaching of mathematics. The availability of user-friendly dynamical software programs has paved the way for a radical yet smooth way for changing the way mathematical concepts are perceived. This chapter presents some of the author's attempts for employing innovative methods in teaching topics in calculus, in differential and difference equations. The focus is on the use of dynamical programs that boost the visual component of the topics being investigated, hence contributing to a more complete understanding of these topics.

INTRODUCTION

In the year 2000, the Standards of the National Council of Teachers of Mathematics (NCTM, 2000) acknowledged that multiple representations help students solve problems, support their understanding of concepts, and better communicate mathematical ideas. Consequently, mathematical visualization has become a core

DOI: 10.4018/978-1-4666-4050-4.ch004

element for learning mathematical concepts. The advancement of technology has largely reshaped the teaching and learning of mathematics. It is now widely accepted that the strength of technology is mainly in its capability for providing greater and easier access to multiple representations of concepts (Kaput, 1992; Porzio, 1999), in particular visualizing mathematical concepts and ideas, and even sometimes abstract theorems (Pool, 1992).This chapter is intended on one hand to present some ways to exploit the dynamical feature of few mathematical programs so as to introduce and clarify concepts in calculus and in an introductory differential equations course. On the other hand, dynamical programs can also be used to conjecture and write proofs and thus may also be useful for research purposes. The chapter ends with an example showing how a dynamical software program created for a specific purpose can help conjecture new results for linear iterative systems.

BACKGROUND

Emphasizing the importance of mathematical visualization can be traced back to the early 1990s. The Mathematical Association of America published a note in 1991 on visualization in mathematics (Zimmermann, 1991). In this publication, Zimmermann, the Note editor, wrote: "Conceptually, the role of visual thinking is so fundamental to the understanding of calculus that it is difficult to imagine a successful calculus course which does not emphasize the visual elements of the subject" (p. 136). Even though Zimmermann targeted only calculus, research and practice have shown that visualization can be critical not only to add a new dimension to many topics in various fields of mathematics, but is also a mean for conjecturing and for writing proofs. The emphasis on visualization began at a time when technology was becoming more widely used in education; however, the technology employed was still more or less static except for few attempts. Reformers were using Maple, Mathematica but others were thinking ahead trying to explore the full power of technology. One cannot but refer to a pioneering dynamical software program called MacMath (see the reference to this software in other chapters of this book) developed at Cornell University by John Hubbard and Beverly West (authors of two other chapters in this book). With dynamic mathematical software programs, the benefits of a visual representation are doubled to say the least. According to Noss and Hoyles (1996, p. 245), when the algebraic representation is complemented with a dynamic environment, it "comes alive". Such an environment therefore constitutes "a rich experimental arena: Students receive the feedback as a direct consequence of their actions and not as a judgmental statement from their teachers" (Arcavi, A. 2008). In addition, one should not undermine the role of the instructor in properly using the technology in the classroom. At a time marked with tremendous advances of educational technological aids (be it Computer Algebra Systems (CAS) which are becoming also dynamical, and specialized dynamical software programs), the role of the teacher is becoming more and more critical as the primary user of technology in the classroom. For many researchers, technology provides a setting for conjecture and creativity for students and teachers alike. The issue however is not as simple as it sounds. Monaghan (2004) writes: "teaching with digital tools does not simply mean considering the software and hardware used." As a matter of fact, even though the young generation of teachers has a positive perception of technology yet in many a case, teachers deliver their classes in almost the same way they were initially taught. And even when they use technological aids, teachers have some difficult decisions to make such as how to decide on the software programs to be used and most importantly how to carefully design student tasks that respond to the needs of the lesson (Laborde, 2001) and benefit from the technology employed. As per Blyth and Labovic (2009), the "development of innovative teaching materials which exploit the strengths of the CAS for tightly integrated eLearning and eAssesment ..., can be used to produce active and engaged learners of mathematics and the technology".

In the sections that follow, examples of how dynamical software programs may be used to enhance learning are presented. Various such program were chosen, some are open source and specialized (e.g. *Interactive Differential Equations*, *ODE Architect*, and *Linear Iterative Systems*), while others are computer algebra systems (such as *Mathematica*). Choosing these programs were based on criteria such as being dynamic, user friendly, useful for the particular topic being discussed, familiarity, and accessibility to students. In the case of Interactive Differential Equations, the creators of the program have provided the user with a large list of dynamic tools which are topic specific. Many of these tools are useful for topics I teach in differential equations and some of them are innovative and creative. My role thus was to choose the ones that fit best my objectives in the course. In the case of Mathematica, the latest versions of the software have become dynamic through an animation slider that one can create. In addition, online projects provided by the Wolfram Research, the makers of Mathematica, are available when you purchase the software. In the case of the Calculus classes I usually teach, I use many of these online aids and in some cases, I introduce some changes to suit my purposes. In the case of the Linear Iterative Systems program, it has been developed upon my request and therefore it already suited my purposes for teaching, learning, and research. All of the prior will become clear as we run through some of the examples in the next section. These examples have been chosen because for their creativity, their usefulness, and their impact on learning.

GRAPHING POLAR FUNCTIONS

In a college level calculus course, the topic of polar coordinates is introduced and discussed in somehow great details. A considerable amount of time is spent graphing polar curves. My rather long experience in teaching has proven that this topic is challenging for the students because graphing is done in the Cartesian plane yet points are plotted based on their polar coordinates. Students have always sketched curves of the form $f\left(x,y\right)=c$ and as such the sudden switch to sketching $r=f\left(\theta\right)$ in the same Cartesian plane is thought-provoking. In general, calculus textbooks (and instructors as well) introduce two sketching strategies to the learners: One is straightforward and suggests that the sketch be done based on a list of points $\left(r,\theta\right)$; the other proposes sketching the corresponding Cartesian curve and then translating the information into the language of polar coordinates. The latter strategy is more mathematically correct since the corresponding *xy*-curve describes completely and accurately how the dependent variable *y* depends on *x*. Consequently the equivalent polar curve is as complete and accurate. In practice however, students usually face tremendous difficulties moving back and forth between the Cartesian plot and the polar one. The main challenge is that a change in *x* (which is a horizontal motion) corresponds to a change in θ (which is a rotational motion) while a change in *y* (vertical motion) corresponds to a change in *r* (a non-defined motion that depends on θ). Here is an example.

Example

If asked to plot the curve $r=\sin\left(2\theta\right)$ also known as a 4-leaf rose, the student is expected to plot first $y=\sin\left(2x\right)$ say over the interval $\left[0,2\pi\right]$ (See Figure 1). The sketch and some analysis of the equation combined produce the conclusions in Table 1.

Figure 1. The plot of $y=\sin\left(2x\right)$ *and* $r=\sin\left(2\theta\right)$ *as generated by Mathematica 8*

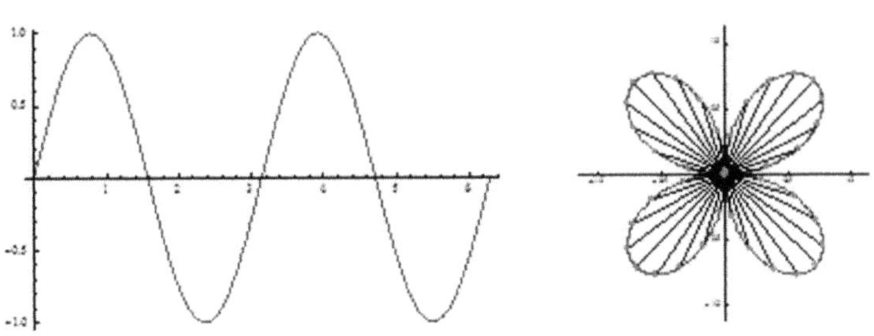

Table 1. Conclusions

$\left[0, \frac{\pi}{4}\right]$	Function increases from 0 to 1	Function is positive
$\left[\frac{\pi}{4}, \frac{\pi}{2}\right]$	Function decreases from 1 to 0	Function is positive
$\left[\frac{\pi}{2}, \frac{3\pi}{4}\right]$	Function decreases from 0 to -1	Function is negative
$\left[\frac{3\pi}{4}, \pi\right]$	Function increases from -1 to 0	Function is negative
$\left[\pi, \frac{5\pi}{4}\right]$	Function increases from 0 to 1	Function is positive
$\left[\frac{5\pi}{4}, \frac{3\pi}{2}\right]$	Function decreases from 1 to 0	Function is positive
$\left[\frac{3\pi}{2}, \frac{7\pi}{4}\right]$	Function decreases from 0 to -1	Function is negative
$\left[\frac{7\pi}{4}, 2\pi\right]$	Function increases from -1 to 0	Function is negative

In terms of polar coordinates, the x - intervals correspond to a range of the θ values; the decrease or increase of the function corresponds to a decrease or decrease of r as the values of θ change; and the sign of the function corresponds to the region in which the sketch is to be plotted. More precisely, when r is negative for a given θ, the point plotted is $\left(-r, \pi + \theta\right)$. It is in particular these details that are mind boggling for the students. Still (non-dynamic) technological aids will not assist in better understanding how the plot is done (See Figure 1). They provide only a teaching aid for the instructor. With a dynamical software program, the plot comes alive (See Figure 2). *Mathematica 8* has been used plot the sketches. Of course there are many other software programs that can do the same job and the reader is welcome to use any other software program that he/she is familiar with.

Figure 2. Snapshots of the polar curve $r = \sin(2\theta)$ *generated by Mathematica 8*

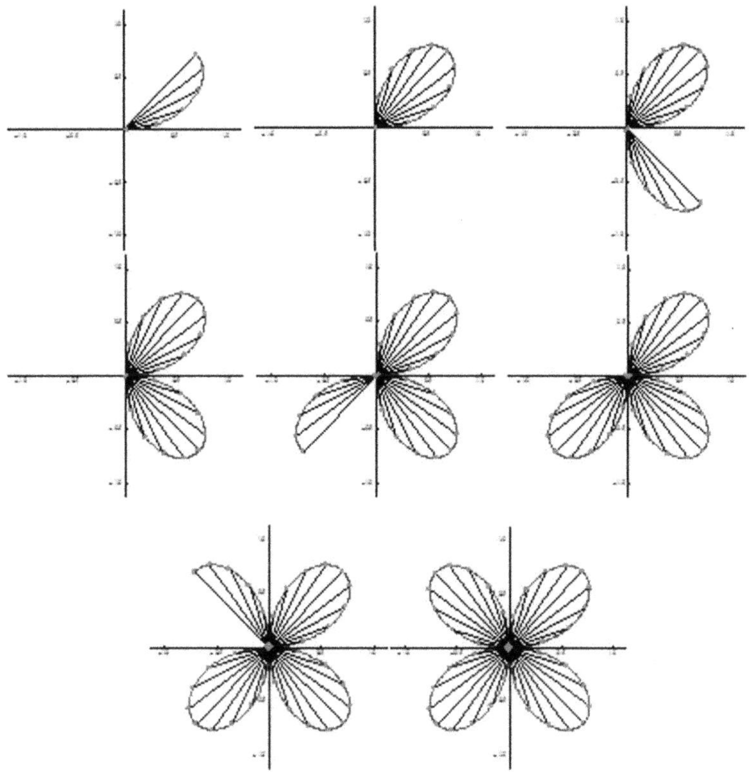

When seeing only the final outcome as a still picture, different students may think differently about the way the rose is plotted (for example, students might think that each leaf is drawn completely and separately in each of the four quadrants of the plane; consequently they might also students conjecture about the order of the leaves). However viewing dynamically how the "leaves" are completed and the order in which they are plotted almost always generates enthusiasm among the learners. Figures 1 and 2 and Figure 3 are samples of 4-leaf roses obtained from two different polar curves: $r = \sin(2\theta)$ and $r = \cos(2\theta)$; yet, the plot is done in different ways. On one hand and in the case of $r = \sin(2\theta)$, the leaves are completed one at a time but they are done in the following order: 1st quadrant, 4th quadrant, 3rd quadrant, and finally the 2nd quadrant. On the other hand, the leaves $r = \cos(2\theta)$ are not completed one at a time, rather the motion of the plot moves from the first quadrant (half a leaf), to the third, the fourth (a full leaf now), to the second, back to the third (another full leaf), to the first, the second (a third full leaf),

Figure 3. The plot of $r = \cos\left(2\theta\right)$ *as generated by Mathematica 8*

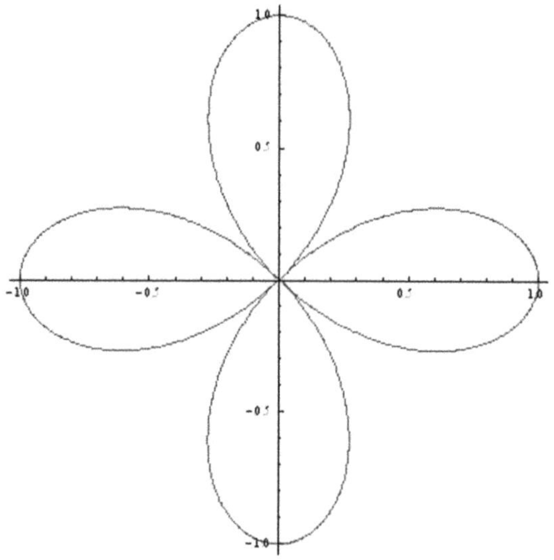

and finally to the fourth (completing the fourth leaf). Many a time, the students themselves will ask to change the function in order to observe more live polar plots. As a teacher, one can exploit this interest and expand the discussion to issues that may not be part of the regular curriculum but that require critical thinking, hence widening the students' horizons: Functions can be changed; the arguments can also change dynamically, so is the number of plotted points (See Figure 4 of the plot of the curve $r = \cos\left(3\theta\right)$ where the argument has been changed). Various questions arise for students to investigate such as how the number of leaves is related to n (whereas $r = \sin\left(2\theta\right)$ and $r = \cos\left(2\theta\right)$ form 4-leaf roses, $r = \cos\left(3\theta\right)$ forms a 3-leaf rose); what distinguishes the rose generated by a cosine function from the one generated by the sine function; what interval is needed (and why) in the Cartesian plot to generate the entire polar plot.

In conclusion, seeing the live plots contribute to a greater understanding of the connection between the Cartesian and polar plots; it also generates as many questions as answers, thus contributing to an expansion of the students' critical thinking. The dynamical feature of the plots can further be exploited if one is able to provide the students with a tool that shows simultaneously the two plots and to show in details the association between the variables in the Cartesian plot (x and y) and the variables in the polar plot (r and θ). It is an idea to be pursued and I hope that software developers take it up for further examination and follow-up.

Figure 4. The plot of $r = \cos\left(3\theta\right)$ *as generated by Mathematica 8*

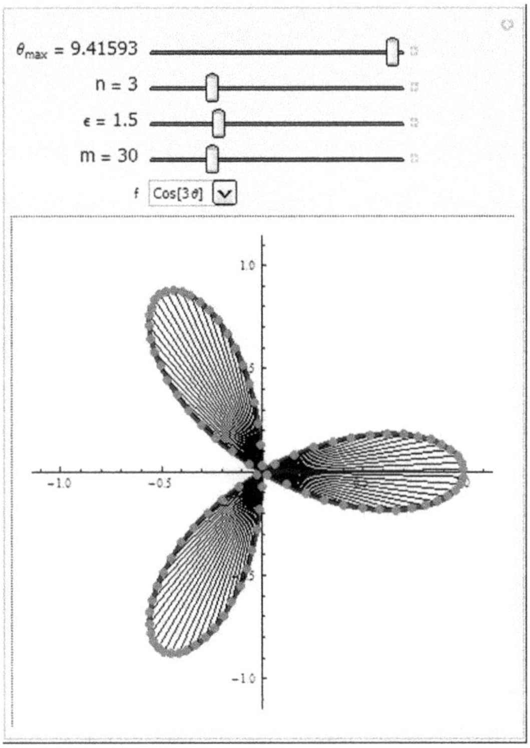

PHASE PORTRAITS FOR LINEAR SYSTEMS

A “modern” introductory course in differential equations emphasizes a qualitative approach for learning the material. An emphasis is placed on visualizing solutions and analyzing them rather than solving equations algebraically and trying to find the solutions in a closed form format. For first-order differential equations, the recipe-type approach for solving quantitatively is marginalized and replaced by an approach that requires sketching solutions mainly by means of a direction field. Although the emphasis in this chapter is on second order equations, yet I wish first to begin by highlighting the importance of visualization tools for understanding the solutions of first order differential equations.

A first-order differential equation takes the form $\frac{dy}{dt} = f\left(t, y\right)$. At each point of the *ty*-plane, the function $f(t, y)$ determines the slope and hence the equation of the tangent line to the solution curve passing through that point. A plot of many of the (mini) tangents is called the *direction field*. The field provides a visual approach for solving the differential equation: Beginning at any point, one follows the directions of the (mini) tangents to obtain a polygonal line that approximates the solution. The more (mini) tangents are drawn, the smoother the polygonal line is. As such, specialized software programs provide learners with tools for drawing the direction field and sketching solutions to initial value problems. This visual approach replaces the recipe-type approach that works only for few types of equations. Figure 5 is the direction field of $\frac{dy}{dt} = y - t$ (generated with a dynamical software program called ODE Architect).

Although visualization has proven to be a key element for understanding solutions to first-order differential equations regardless of the equation type, it is not very important for the visualization tools to be dynamic. This is not the case though for second order differential equations.

A Second-order differential equation takes the form $\frac{d^2y}{dt^2} = f\left(t, y\right)$. In a reformed setting, such equations are discussed after being transformed into linear systems of first-order differential equations as follows: Define a new variable $v(t)$ such that $\frac{dy}{dt} = v(t)$, then $\frac{dv}{dt} = f(t, y)$. We now have a system of two first order differential equations; a *vector field* in the *yv*-plane is constructed where at every point a vector is sketched with components $\frac{dy}{dt}$ and $\frac{dv}{dt}$. Starting at any given point of the *yv*-plane, if one follows the vectors of the vector field, a *solution curve* is obtained.

Figure 5. The direction field of the ordinary differential equation $\frac{dy}{dt} = y - t$ *as generated with ODE architect*

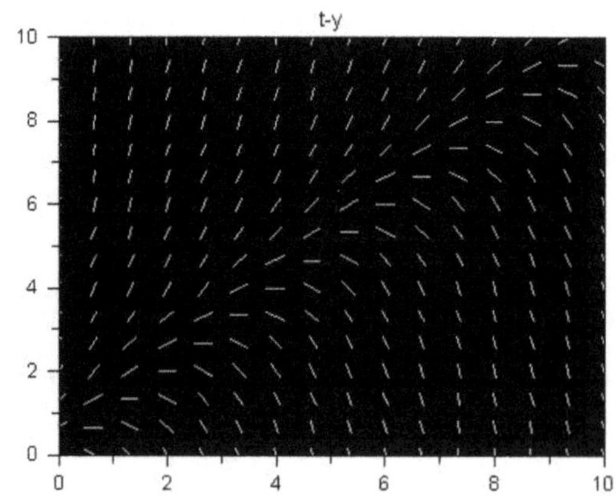

Whereas in the case of a first-order differential equation, it is clear that solutions are drawn from left to right as time *t* increases, the motion of a solution curve for a system of first-order differential may not be as obvious since the independent variable *t* does not show in vector fields. Thus, by viewing solution curves as they are being plotted, the dynamic sketch specifies the direction of motion on the solution curve and hence students understand further how to use the vectors of the vector field when drawing such curves. In addition, an understanding of the system solution curves contributes to an understanding of the separate solutions $y(t)$ and $v(t)$. In particular, the learner is able to investigate the long term behavior of these solutions, something almost inconceivable in the traditional approach. While such an approach in the past was not popular because of the difficulties associated with the visual aspect, the advancement of computer graphics particularly the development of specialized dynamical software programs has provided teachers with unprecedented visual means to explore dynamically second-order differential equations.

The relative ease for using the graphical approach in such situations (due to the availability of software programs) has always prompted me to begin the discussion of systems with a real life example, generally known as the Predator-Prey system, even before introducing the math behind the problem. The next example describes the population of a predator versus that of a prey. Starting with a pair of populations, the student can view dynamically how the two populations vary accordingly (See Figure 6). Students are usually asked to derive a rough sketch of the population versus time of each species separately.

Example

Consider the following Predator-Prey system, where x represents the predator and y the prey:

$$\begin{cases} \dfrac{dx}{dt} = -0.5x + \dfrac{xy}{3} \\ \dfrac{dy}{dt} = \left(1 - y\right)y - \dfrac{xy}{0.3} \end{cases}$$

Figure 6 is one particular solution curve for this system.

In this example, students are lead to conclude that the two populations reach an equilibrium state. Students are then curious to investigate different cases. Examples where species may become extinct or examples where populations are cyclic are investigated graphically without any mathematical background.

Figure 6. The vectorfield of a predator-prey system together with one solution curve

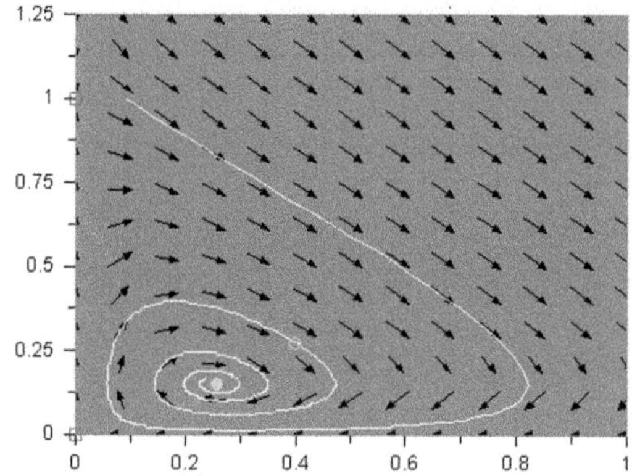

These discussions are then followed of course by a detailed discussion of systems and in particular linear systems of differential equations. These take the form:

$$\frac{dx}{dt} = ax + by; \frac{dy}{dt} = cx + dy$$

In all such systems $\left(0,0\right)$ is a solution called the *equilibrium* solution. A complete understanding of such systems requires an analysis of the coefficient matrix $\begin{bmatrix} a & b \\ c & d \end{bmatrix}$. More precisely, one has to analyze the characteristic equation of the matrix $\lambda^2 - 2\mathrm{tr}(A)\lambda + \det(A) = 0$, where $\mathrm{tr}(A)$ and $\det(A)$ are the trace and determinant of *A* respectively. The shape of the solution curve in the *xy*-plane depends on the type (real vs. complex) and number of the eigenvalues and also on the corresponding eigenvectors. One specialized interactive and dynamical open access software that has been available in the market for over a decade now is Interactive Differential Equations (IDE: http://www.aw-bc.com/ide/). IDE is designed by Hu Hohn who serves as the director of Computer Arts at Massachusetts College of Art in Boston. The program is an open ended-solver and is a collection of 92 labs built each to offer a complete understanding of a particular concept in ordinary differential equations. These labs were carefully designed by mathematicians that were pioneers in leading the reform efforts in ordinary differential equations (such as author Beverly West who is also a chapter author, and Jean-Marie McDill, one of the book editors). For instance, one is able to view the changes in the phase portrait of a linear system in a dynamical fashion. Figures 7a, 7b, and 7c, generated with IDE, offer a dynamic visual alternative to the linear classification of the equilibrium solution $(0,0)$ for the 2×2 linear systems of differential equations (source, sink, saddle, spiral source, spiral sink, or center). The user is able to manipulate the matrix coefficients $a, b, c,$ and d producing a dynamic environment and allowing the student to see in real time how for instance straight-line solutions can collapse to become one and how the one straight-line solution disappears in favor of solution curves that spiral around the origin (the equilibrium solution). The user also observes the changes in the behavior of all other solutions (for instance how spirals change to become closed loops). In such systems, one is also interested in investigating the time-series of the solutions, that is the corresponding *ty* and *tv*-sketches since the *tv*-time series are the actual solutions of the original differential equation $\frac{d^2y}{dt^2} = f(t, y)$.

Figure 7. (a) The phase portrait of a linear system with two straight line solutions generated using IDE; (b) the phase portrait of a linear system with solutions spiraling around the origin; (c) the phase portrait of a linear system with solutions forming closed loops around the origin

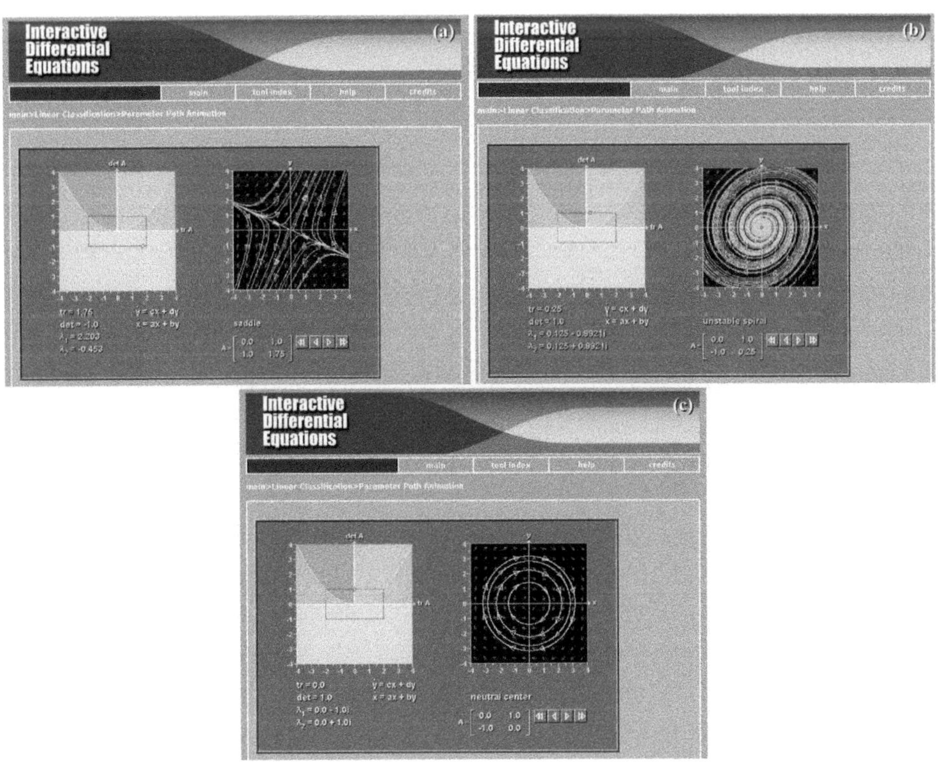

Instructors can easily exploit the dynamical nature of the program and request from students an analysis of the different time series (shapes and long-term behavior). As was mentioned earlier, real-life examples can largely contribute to a better understanding of systems. In the linear case, one can investigate the mass-spring problem; one can also discuss the motion of a door (closing smoothly vs. non-smoothly, and also vs. an oscillating door). The developers of IDE have thought of yet another application which may not be found in other textbooks/technological tools; they have called the it "Romeo and Juliet". This dynamical example describes a love affair between Romeo and Juliet using a system of first order differential equations. In this tool x denotes Romeo's love for Juliet and y denotes Juliet's love for Romeo (both changing with time). By changing the values of parameters a, b, c, and d, one can explore various possible situations and thus students may be asked questions like:

1. Look at the phase plane and time series for x and y to analyze the ensuing relationship. How would you characterize their relationship?
2. Examine the time series for x and y. What are Juliet's feelings for Romeo when he is most attracted to her?

Here is an example that is relatively easy to investigate.

Example

Consider the case where a is set to be 0, b = -0.2; c = 0.8, and d = 0:

$$\begin{cases} \dfrac{dx}{dt} = -0.2y \\ \dfrac{dy}{dt} = 0.8y \end{cases}$$

The phase portrait is a closed loop (See Figure 8) and the-time series are oscillating periodic functions. As such, students can easily provide answers to the previous two questions.

More complex cases can of course be considered. As such, reading the time series and relating it to a solution curve in the phase plane can be a challenging skill for a student. Viewing the two graphs plotted simultaneously can improve this skill. Thus, using real-life examples such as a love affair between Romeo and Juliette will certainly engage more the students contributing to a better understanding of linear systems of differential equations.

Figure 8. Screen shot of the Romeo and Juliet tool

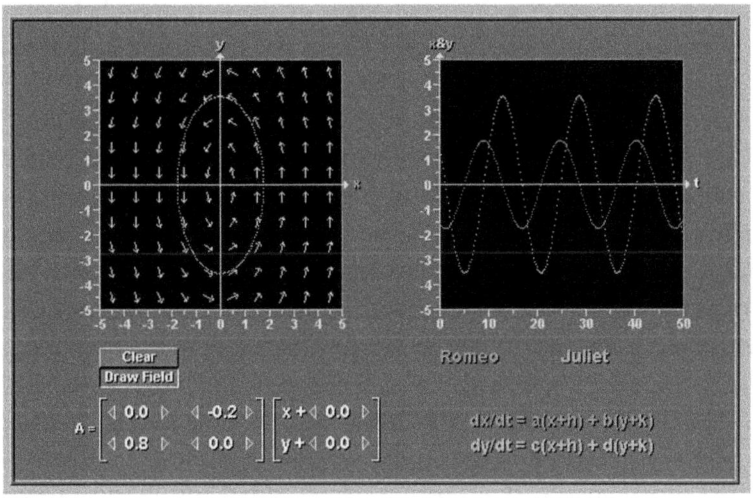

BEYOND THE TRADITIONAL CURRICULUM: LINEAR ITERATIVE SYSTEMS

Rarely are discrete dynamical systems discussed at the undergraduate level even though the treatment of such systems is in many ways similar to systems of difference equations. Linear iterative systems take the form:

$$\begin{cases} x_{n+1} = ax_n + by_n \\ y_{n+1} = cx_n + dy_n \end{cases}$$

As in thc casc of linear systems of differential equations, $(0,0)$ is an equilibrium solution; other solutions take the form $\vec{Y}_n = \lambda^n \vec{v}$ where λ and $\vec{v}$ denote an eigenvalue and a corresponding eigenvector for the system coefficient matrix $A = \begin{bmatrix} a & b \\ c & d \end{bmatrix}$, very much like the case of linear systems of first order differential equations. In case *A* has two distinct eigenvalues (real and complex) the general solution takes the form $\vec{Y}_n = \lambda_1^n \vec{v}_1 + \lambda_2^n \vec{v}_2$. Other possibilities also arise: One double eigenvalue and two complex conjugates eigenvalues. These cases are discussed in details in Habre and McDill (2008). The shape and long-term behavior of the solutions to the iterative system clearly depends on the number of eigenvalues and on their sign in case they are real. The analytic form of the solution however may not speak much about neither the shape or the long-term behavior; for this reason a technological tool that can add a visual and dynamical insight to the solutions obtained seemed to be necessary The IDE tool developed by Hu Hohn for the classification of solutions in systems of first order differential equations lead the author to ask the designer to develop a similar tool for 2×2 linear iterative systems (that we called *Iterated Linear Systems*). The tool is not as sophisticated as the one for differential equations because it was not meant to be commercialized. However it does serve the intended purposes. To be more specific, the window of the tool consists of a system matrix, its eigenvalues and eigenlines, the trace and determinant of the matrix, the trace-determinant plane, and the phase plane with or without the vector field. The latter object adds a visual dynamic dimension to the solution behavior as time progresses allowing an investigation of the long-term behavior and consequently a classification of the origin. As in the continuous tool, a manipulation of the matrix coefficients *a, b, c,* and *d* produces a dynamic environment allowing users to observe in real time the change in the solution behavior. The next examples are 2 illustrations of solution behaviors and consequently various classifications of the equilibrium solution $\left(0,0\right)$.

Example

Consider the iterative linear system:

$$\begin{cases} x_{n+1} = 1.05y_n \\ y_{n+1} = -1.1x_n \end{cases}$$

The eigenvalues of the system are pure imaginary $\lambda = \pm i\sqrt{1.155} \approx \pm 1.075i$. The general real solution of the system takes the form:

$$\vec{Y}_n = c_1\left(1.075\right)^n \begin{pmatrix} \cos\left(n\dfrac{\pi}{2}\right) \\ -\dfrac{1.075}{1.05}\sin\left(n\dfrac{\pi}{2}\right) \end{pmatrix}$$

$$+c_2\left(1.075\right)^n \begin{pmatrix} \sin\left(n\dfrac{\pi}{2}\right) \\ -\dfrac{1.075}{1.05}\cos\left(n\dfrac{\pi}{2}\right) \end{pmatrix}$$

The qualitative behavior of such a solution may not be clear unless it is complemented by a figure. Figure 9a is the sketch of one solution. The dynamic nature of the program allows the user to observe that the solution *spirals out* away from the origin. We call the origin in this case a *spiral source*. Next is another example exhibiting a different behavior.

Figure 9. (a) A window showing that the origin is a "spiral source" (b) a window showing the case of a repeated eigenvalue

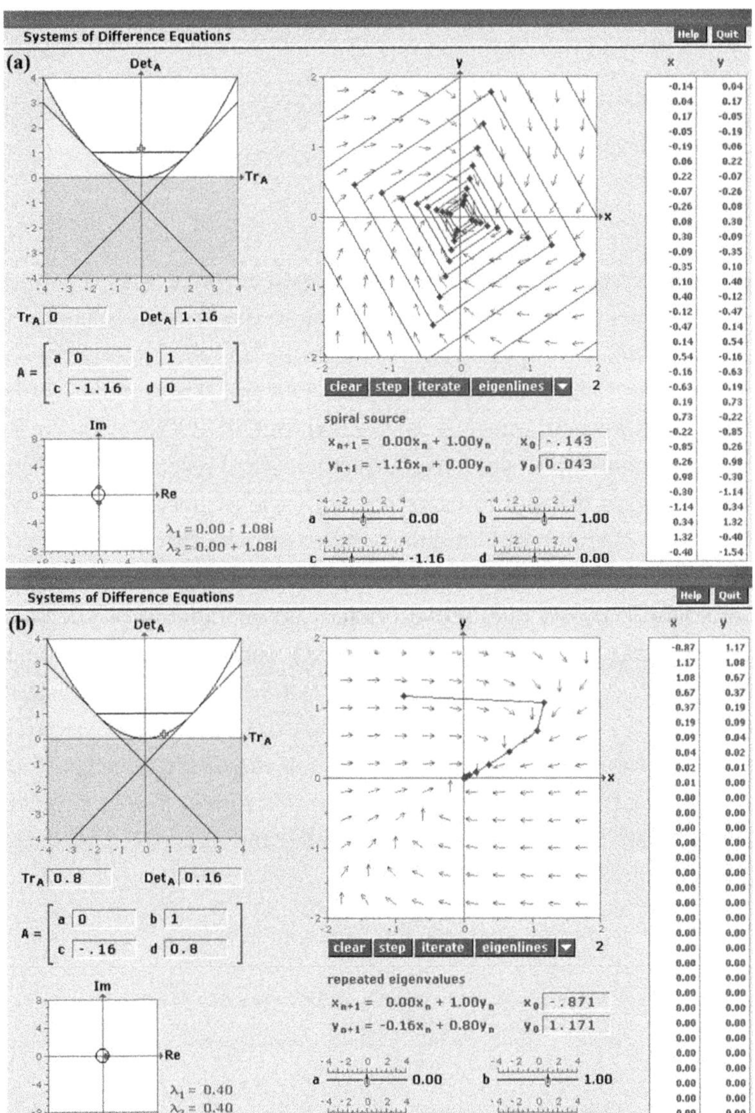

Example

Consider now the linear iterative system with a repeated eigenvalue $\lambda = 0.4$

$$x_{n+1} = y_n; y_{n+1} = -.16x_n + 0.8y_n$$

In a research conducted at the time of the development of the software, the author in collaboration with Jean-Marie McDill from California Polytechnic State University proved that the general solution in this case takes the form (Habre and McDill, 2008):

$$\vec{Y}_n = c_1\lambda^n\vec{v} + c_2\lambda^n\left(n\vec{v} + \vec{u}\right)$$
$$\text{where } \left(A - \lambda I\right)\vec{u} = \vec{v}$$

Analyzing the long term behavior of this solution is not obvious but the use of the tool provided a visualization of the solution (See Figure 9b).

The tool has thus provided a new dimension to the previous examples. On the one hand, its dynamical property that results from "moving" along the Trace-Determinant plane has assisted us (Jean-Marie McDill and I) in *completely* classifying all 2×2 linear iterative systems (see Habre and McDill, 2008). As such, the development of this dynamic tool was a useful mathematical research tool. On the other hand, one can exploit the tool to introduce the topic of linear iterative systems to the undergraduate curriculum particularly in an introductory differential equation course. The treatment of linear iterative systems is similar to the linear systems of differential equations. The qualitative (visual) understanding of such systems is enhanced if dynamical tools are used. Similarities and differences can be investigated and I find it only natural that the two types of systems be discussed in a classroom setting. Details of the analytical properties of iterative systems need not be the focus (the form of the general solutions in the different cases of the trace-determinant plane), rather the dynamic tool *Iterated Linear Systems* can be used to discuss the qualitative properties of the solutions and hence to a classification of the equilibrium solution.

CONCLUSION

In conclusion, dynamical mathematical software programs can be used in the classroom to contribute to an improvement of the learning process and to add new dimensions to topics that are part of a traditional mathematics curriculum (e.g. calculus and differential equations). It can also be used in research (such as in the case of the classification of the origin in linear iterative systems). The availability of many such programs in the market (as well as the many open access tools) can be confusing as to what to use and how to choose the appropriate ones. This is indeed an important question and users (especially educators) have to clearly identify their own purposes for incorporating such tools into the learning process. Equally important is how the software program is used once it is chosen. Teachers have to carefully design tasks that are in line with the learning outcomes intended for using these technological tools. These questions and issues were not the focus of this chapter but they are worthy of contemplation and further research in order to close the learning loop.

REFERENCES

Abboud-Blanchard, M. (2005). Uses of ICT by pre-service teachers. In F. Olivero & R. Sutherland (Eds.), *Proceedings of the 7th International Conference of Technology in Mathematics Teaching*, (Vol. 2, pp. 74 - 78). Technology in Mathematics Teaching.

Arcavi, A. (2008). Modeling with graphical representations. *For the Learning of Mathematics*, *28*(2), 2–10.

Blyth, B., & Labovic, A. (2009). Using maple to implement elearning integrated with computer and assessment. *International Journal of Mathematical Education in Science and Technology*, *40*(7), 975–988. doi:10.1080/00207390903226856.

Eisenberg, T., & Dreyfus, T. (1991). On visual versus analytical thinking in mathematics. I C. Hoyles, R. Noss, & R. Sutherland (Eds.), *Proceedings of the Tenth International Conference on the Psychology of Mathematics Education*, (vol. 1, pp. 153-158). Psychology of Mathematics Education.

Habre, S., & McDill, J. (2008). A complete classification of 2×2 linear iterative systems. *The Electronic Journal of Mathematics and Technology, 2*(1).

Hohn, H. (2005). *Iterated linear systems.* Unpublished.

Hohn, H., West, B. H., Strogatz, S., McDill, J. M., & Cantwell, J. (2008). *Interactive differential equations.* Retrieved from http://www.aw-bc.com/ide

Hubbard, J. (1994). What it means to understand a differential equation. *The College Mathematics Journal*, *25*(5), 372–384. doi:10.2307/2687502.

Hubbard, J., & West, B. (1993). *MacMath*. New York: Springer-Verlag.

Hubbard, J. H., & West, B. H. (1992). *MacMath 9.2*. New York: Springer-Verlag.

Kaput, J. (1992). Technology and mathematics education. In Grouws, D. A. (Ed.), *Handbook of Research on Mathematics Teaching and Learning* (pp. 515–556). New York: Macmillan.

Laborde, C. (2001). Integration of technology in the design of geometry tasks with Cabri-geometry. *International Journal of Computers for Mathematical Learning*, *6*(3), 283–317. doi:10.1023/A:1013309728825.

Monaghan, J. (2004). Teachers' activities in technology-based mathematics lessons. *International Journal of Computers for Mathematical Learning*, *9*, 327–357. doi:10.1007/s10758-004-3467-6.

Noss, R., & Hoyles, C. (1996). *Windows on mathematical meanings: Learning cultures and computers*. Dordrecht, The Netherlands: Kluwer. doi:10.1007/978-94-009-1696-8.

ODE Architect. (2010). *Consortium for ordinary differential equations*. New York: John Wiley & Sons, Inc.

Zimmermann, W. (1991). Visual thinking in calculus. In Zimmermann, W., & Cunningham, S. (Eds.), *Visualization in teaching and learning mathematics* (pp. 127–137). Washington, DC: Mathematical Association of America.

Chapter 5
Nonlinear is Essential, Linearization is Not Enough, Visualization is Absolutely Necessary

Beverly West
Cornell University, USA

ABSTRACT

Linear differential equations have been well understood for some time and are an important tool for studying the nonlinear systems that most frequently arise in mathematical models of real world systems. Nonlinear systems do not usually have formula solutions, but with graphics, we can see the behaviors of the solutions and thereby "understand" the differential equations. Dynamic and interactive presentations provide students with a streamlined route to understanding these behaviors, resulting in immense power and efficiency that was not previously available at the undergraduate level.

INTRODUCTION

Mathematical models of real world problems are almost always nonlinear. We can linearize many such models to give approximate solutions, but these approximations are only good in a sufficiently small neighborhood of an equilibrium point. However, we show that often what at first glance appears to be an appropriately small neighborhood is not at all sufficient for accurate long-term prediction, which is usually the objective.

DOI: 10.4018/978-1-4666-4050-4.ch005

Our examples will be from Differential Equations (DEs), which provide a vital connection between mathematics and the real world. We proceed from the most elementary to more advanced examples, showing how dynamic graphic software helps students to see the behavior, to know when and where to trust a linear approximation, as well as what to do when you cannot. We get the basic information in two-dimensional systems from the *xy* phase plane representation, but students need to understand the connection to the solution graphs, *tx* and *ty*. We also sketch advantages and issues of graphic software for higher order differential equations.

What is of utmost importance for DEs is not a formula for the solutions, which usually does not even exist for a nonlinear DE, but rather the *behaviors* of the solutions, which are easily absorbed through their pictures (created by numerical methods). We must attend, of course, to the issue of when can we trust such pictures.

Furthermore, we need different kinds of dynamic graphic software. First, to convey concepts, we have found immense benefit in *interactive* illustrations that provide guided exploration of good examples. The package on which we rely is *Interactive Differential Equations* (Hohn et al, 2000). Second, to realize the *power* that the graphic revolution has unleashed, we need open-ended solvers that allow students to enter their own DEs for graphic experimentation; for this, we have always used *MacMath* (Hubbard & West, 1995, 1999).

We shall begin with the background of DEs with a graphic interface, then review the linear behaviors that are well-understood. Next, with examples, we show the complications that arise in nonlinear DEs, and why linearization about the equilibria does not completely explain the behaviors. We extend the discussion to include chaotic behavior and the assistance provided by software that deals with higher dimensions. Lastly, we discuss the two types of software and what remains to be done for the open-ended version to be useful.

BACKGROUND

Differential Equations and their Graphs

We shall concentrate in this chapter on two-dimensional systems of autonomous differential equations, of the form

$$\frac{dx}{dt} = f(x,y), \qquad \frac{dy}{dt} = g(x,y)\,. \tag{1}$$

This system defines a *vector field,* where at any point in the xy-plane the DE tells the direction the trajectory $\left(x(t), y(t)\right)$ must follow through that point. Furthermore, there are existence and uniqueness theorems[1] that guarantee, for "reasonable" functions f and g, that only *one* solution goes through each point in xy-space.

The xy graph of trajectories is called the *phase portrait.* The individual solution graphs x(t) and y(t) can also be graphed, separately, as *time series* or *solution graphs.*

Graphic Software for Differential Equations

With properly designed dynamic software a student can enter a DE, display the vector field, and then choose an initial condition to start a trajectory at any point in that field simply by clicking on the graph at the desired point. In other words, the click determines the initial conditions by the coordinates of the point of click.

Software draws solutions $x(t)$ and $y(t)$ or trajectories $\left[x(t), y(t)\right]$ step by step, using a numerical method. In general, the smaller the stepsize, the better the accuracy; many programs allow choosing the numerical method.[2]

If the stepsize is small enough, and the graphing speed slow enough, to watch the trajectory evolve, the student can see exactly *how* it follows the vector field. This simple fact is one that many students missed before the age of computer graphics, no matter how hard we tried to explain it in text or lecture. Once they saw a solution *evolve* through the vector field, they immediately grasped this crucial concept, without any formal instruction in that regard. See Figure 1a.

Figure 1. Graphing trajectories and solutions with IDE (Hohn, 2000): (a) constructing solutions step by step; (b) phase portrait, plus graphs of solutions x(t), y(t).

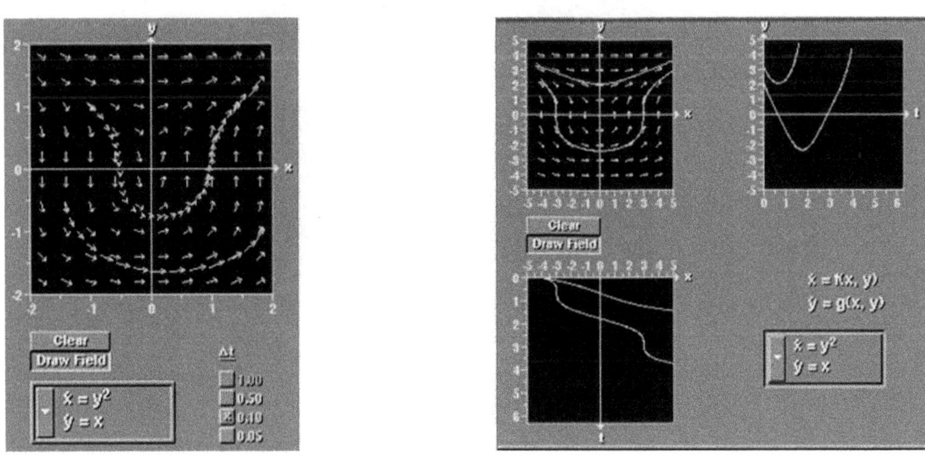

(a) (b)

In this way, we have used *Interactive Differential Equations* (IDE) by Hubert Hohn et al. (2000) to introduce concepts through interactive illustrations. This collection of 96 tools provides guided exploration of selected examples and is available for free on the Internet. Colleagues at MIT developed a similar set of mathlets with Hohn, and have studied carefully the students' experiences (Miller & Upton, 2008). Another such set of interactive illustrations was made by Hohn for the text by Blanchard, Devaney, and Hall (2006).

An improvement to simply seeing the phase portrait is to add the solution curves $x(t)$ and $y(t)$. But static pictures can be difficult for naïve students to interpret. Far more important for student understanding is the ability to see these graphs *evolve*, dynamically and simultaneously. Again we use Hohn (2000) to illustrate.

In Figure 1b, the *tx* graph is oriented so that *x* is in the same direction as on the *xy* phase portrait. The reason for this is that Hohn found that many students understood the time series more readily from this orientation. There is another nice *IDE* tool, Figure 2, which gives even more insight: Students can draw their own curves in the *xy*-plane and watch the $x(t)$ and $y(t)$ graphs evolve simultaneously.

Figure 2. Drawing an xy curve while x(t) and y(t) evolve simultaneously

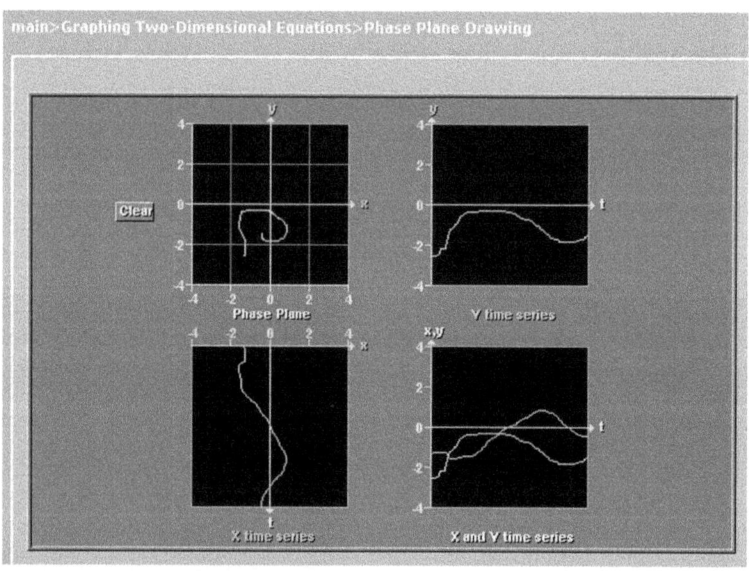

Linear Systems of Differential Equations

Linear systems of differential equations with constant coefficients are those that can be written:

$\vec{x}\,' = A\vec{x}$, where in two dimensions $\vec{x} = \begin{bmatrix} x(t) \\ y(t) \end{bmatrix}$ and $A = \begin{pmatrix} a & b \\ c & d \end{pmatrix}$;

i.e. $\begin{pmatrix} dx/dt \\ dy/dt \end{pmatrix} = \begin{pmatrix} a & b \\ c & b \end{pmatrix}\begin{pmatrix} x \\ y \end{pmatrix} = \begin{pmatrix} ax + by \\ cx + dy \end{pmatrix}$.

They have nice algebraic solutions of the form

$$\vec{x} = \begin{bmatrix} x(t) \\ y(t) \end{bmatrix} = c_1 e^{\lambda_1 t}\vec{v}_1 + c_2 e^{\lambda_2 t}\vec{v}_2,$$

where the c_i are arbitrary constants, the λ_i are the eigenvalues of the matrix A, and the $\vec{v}_i$ are the corresponding eigenvectors.

Linear DEs have simple phase portraits, which can be classified as follows, depending on the eigenvalues and eigenvectors of **A**. For real eigenvalues you can *see* the eigenvectors, something that even surprised and delighted many of our faculty colleagues when such graphics became available in the early 1980s. See Figure 3 for the classification of linear phase portraits.

A dynamic demonstration of this classification is given by *IDE* at http://www.aw- bc.com/ide/idefiles/media/JavaTools/lnclppan.html.

We see in Figure 3 that in general linear DEs have exactly one equilibrium. An exception is when det A = 0, at the boundary between nodes and saddles, where there is an entire line of equilibria. In Figure 4 every point on the x-axis (one of the

Figure 3. Classification of linear equilibria

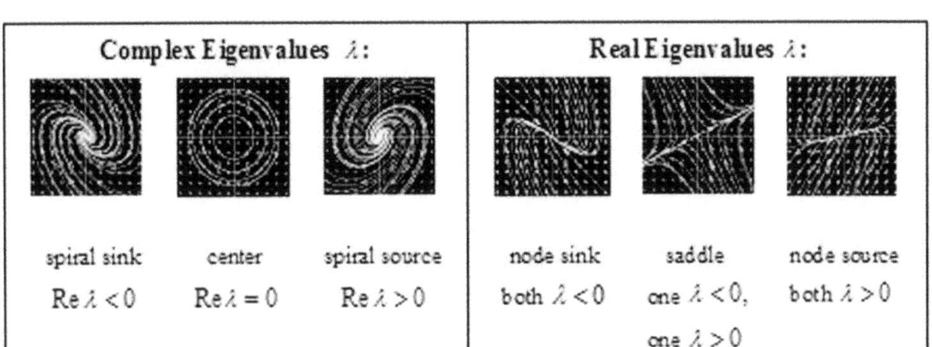

Figure 4. Typical phase portrait at the boundary between nodes and saddles

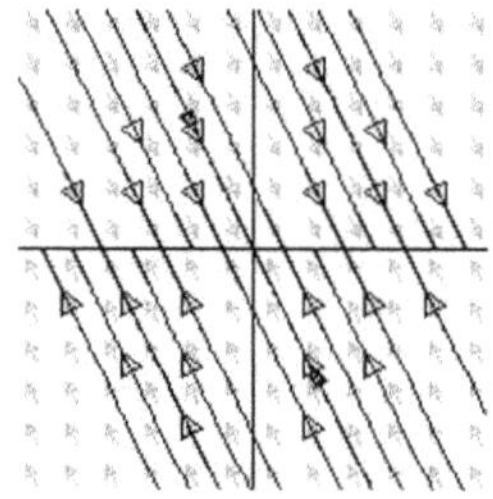

eigenvectors) is an attracting sink, while the other trajectories follow the direction of the other eigenvector.

NONLINEARITY

Comparison with Linear Behaviors

Nonlinear DEs are far less well-behaved than linear DEs. They can have any number of equilibria (including none), and they can also have a periodic behavior showing up in phase portraits as *limit cycles*. Furthermore, nonlinear DEs most often do not *have* algebraic solutions. See, for example, Hubbard and Lundell (2011). Nevertheless, graphing software shows that solutions *do* exist, whether or not one can find a formula for them, once the vector field is defined. Figure 5 shows some examples.

Figure 5. Phase portraits for sample nonlinear systems: (a) a single equilibrium, a spiral source, and an attracting limit cycle; (b) two equilibria – a spiral sink and a saddle – which we shall examine more closely; (c) numerous equilibria and limit cycles, only a few are drawn

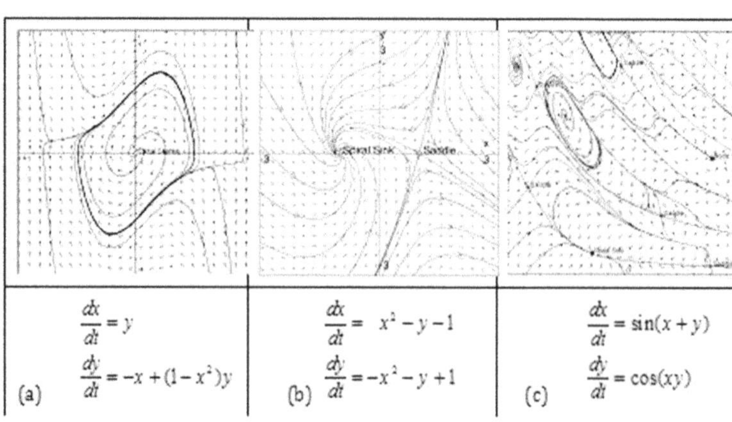

(a)	(b)	(c)
$\frac{dx}{dt} = y$ $\frac{dy}{dt} = -x + (1 - x^2)y$	$\frac{dx}{dt} = x^2 - y - 1$ $\frac{dy}{dt} = -x^2 - y + 1$	$\frac{dx}{dt} = \sin(x + y)$ $\frac{dy}{dt} = \cos(xy)$

Pictorially it is not hard to imagine that zooming in on a nonlinear equilibrium (where $\frac{dx}{dt}$ and $\frac{dy}{dt}$ are zero simultaneously) will eventually look like a linear phase portrait, just as zooming in on a curve $y = f(x)$ at a point $\left(x_0, f\left(x_0\right)\right)$ will eventually look like the tangent to the curve at that point.

The linearization of $\frac{dx}{dt} = f\left(x, y\right)$, $\frac{dy}{dt} = g(t, y)$ at an equilibrium *e* is calculated as follows (Farlow, et al., 2007):

$$\frac{d}{dt}\begin{bmatrix} x - x_e \\ y - y_e \end{bmatrix} = \begin{bmatrix} \frac{\partial f}{\partial x} & \frac{\partial f}{\partial y} \\ \frac{\partial g}{\partial x} & \frac{\partial g}{\partial y} \end{bmatrix}_{(x_e, y_e)} \begin{bmatrix} x - x_e \\ y - y_e \end{bmatrix}. \tag{2}$$

We find for the center example of Figure 5, there are two equilibria, at (-1,0) and (1,0) ; the linearizations about each of them are given in Figures 6a and 6b.

For a linear saddle equilibrium, as shown in Figure 3 (the middle of the real eigenvalue cases) and Figure 6b, the *separatrices* show the vital separation between behaviors. (These are also called *stable* and *unstable manifolds*.) Except for the

Figure 6. Linearizations about each of the equilibria for example Figure 5b. About (-1, 0) the linearization is a spiral sink, and about (1, 0) the linearization is a saddle.

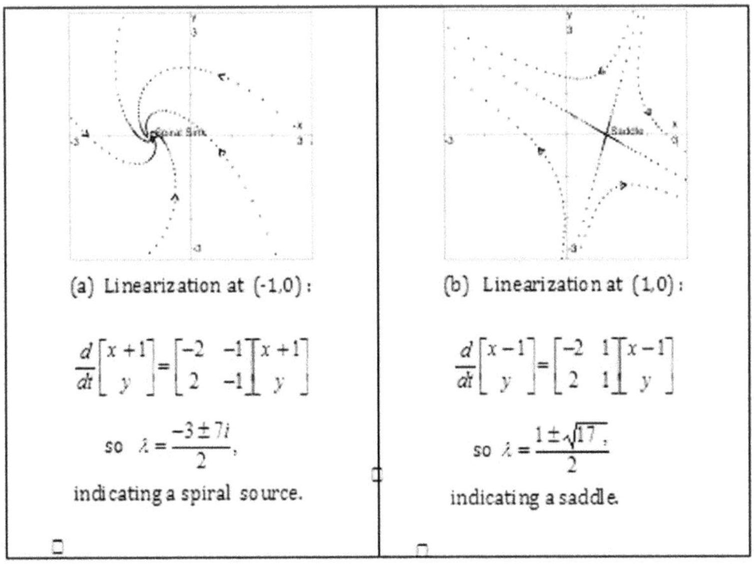

stable separatrices, *all* trajectories in the neighborhood of a linear saddle zoom off to positive or negative infinity, eventually in the directions of the unstable separatrices. The separatrices are the exceptional solutions that start (or end) very close to the equilibria, in the directions of the eigenvectors (you can imagine that the same idea in more than two dimensions, e.g., in a 3-dimensional system you might have a 2D manifold surface defined by two eigenvectors, and a 1D manifold by a third).

Adapting Linearizations for Multiple Equilibria

Look back at the nonlinear phase portrait of Figure 5b. In the linearization of the saddle on the right (See Figure 6b), every trajectory that begins to the left of the unstable separatrices (the more vertical ones) goes off to infinity on the left. However, in the nonlinear phase portrait every trajectory above the unstable separatrices spirals in to the spiral sink on the left, a *totally different behavior*. See Figure 7, where we have shaded the basin of the spiral source, and compare again with the linearizations of Figure 6.

The nonlinearity of the DE in Figure 5b causes the *long-term* behavior to the left of the nonlinear saddle to be *very* different from the linearized saddle; moreover, the nonlinear spirals on the right side of the sink have also been warped from the linearized spiral. Nonlinearity warps the linearizations to "fit together". This is where we need the *pictures* to truly see where the long-term behaviors end up in nonlinear equations.

Figure 7. The nonlinear phase portrait of Figure 5b repeated to shade the basin of the spiral sink

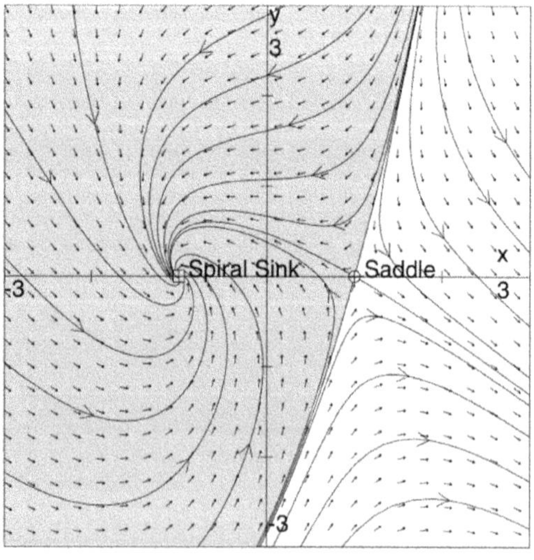

The Need for Exploration and Experiment

Although close to each equilibrium the linear and nonlinear phase portraits look very similar, this difference in behaviors is crucial. The *long-term* behavior is usually what matters in a real world problem; this requires following trajectories *beyond* a small neighborhood of the equilibrium. In our example, we see that the linearization about the saddle is *not* valid in *any* neighborhood to the left of the unstable separatrices. It is essential to realize this when predicting long-term behavior; it is *easy* to see in the nonlinear phase portrait. Hence *seeing* such a phase portrait is essential for a nonlinear system.

A further question arises as to how we might adjust equations to achieve a desired change of behaviors. Coefficients of terms in a differential equation may be treated as changing parameters in the DE. The behaviors change as the parameters are varied, and a whole field of study is directed at bifurcations – the values of parameters at which a dramatic change in phase plane behavior occurs. An excellent example is provided by competition models.

Competition Models

Competition between two biological populations, x and y is often and easily modeled as follows, where the constants a, b, c, d, e, f are all positive:

$$\begin{aligned}\frac{dx}{dt} &= x(a - bx - cy)\\ \frac{dy}{dt} &= y(d - ex - fy)\end{aligned} \tag{3}$$

The rate of change of each population depends first on the population itself, with positive contributions from birth rates (a and d) and negative contributions from crowding and competition with its own species (b and f), and from the other species (c and e).

The relative sizes of these coefficients a, b, c, d, e, f make all the difference in how the populations interact, as many textbooks explore, e.g., Farlow (2002), Hubbard & West (1995). The simplest key to sorting out the possibilities is to locate the *nullclines*, the locus of horizontal or vertical slopes:

v nullclines of vertical slopes (where $dx/dt = 0$, so $x = 0$ or $y = \frac{a}{c} - \frac{b}{c}x$), and

h nullclines of horizontal slopes (where $dy/dt = 0$, so $y = 0$ or $y = \frac{d}{f} - \frac{e}{f}x$).

Nullclines are *not* solutions or trajectories; they are simply the locus of points in the plane where $\frac{dx}{dt} = 0$ or $\frac{dy}{dt} = 0$. Therefore on phase portraits we use dotted or dashed curves for nullclines, to distinguish them from trajectories.

Figure 8 shows for the competition model the four possible orientations of the oblique nullclines. Their intercepts are stated below each graph, with those intercepts on the vertical axis (*a/c* and *d/f*) listed before those on the horizontal axis (*a/b* and *d/e*). In all four cases the axes are also nullclines, though the dotting does not show up – the vertical axis is another ***v*** nullcline, and the horizontal axis is another ***h*** nullcline.

Figure 8. The four possible configurations of nullclines for the competition model

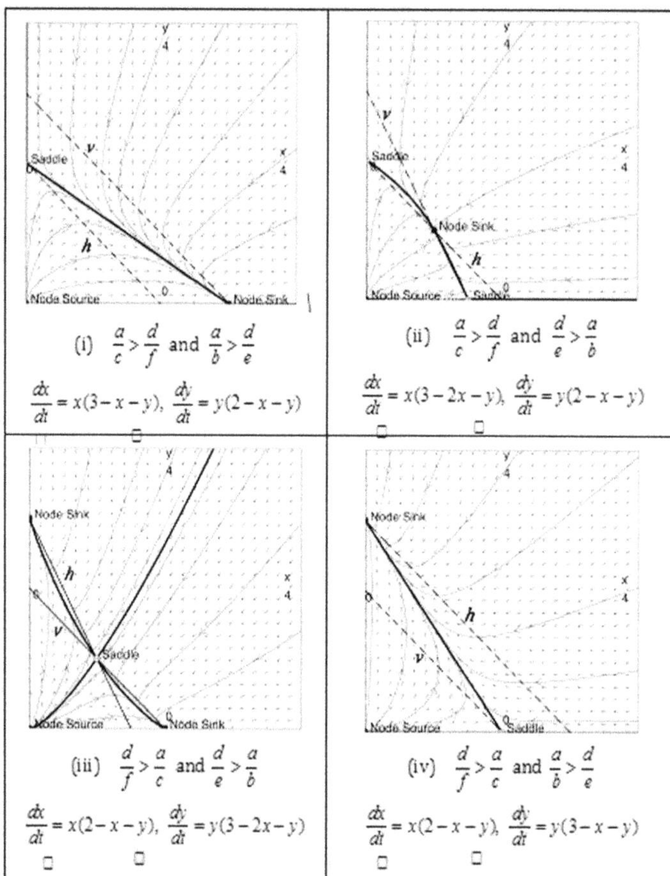

(i) $\frac{a}{c} > \frac{d}{f}$ and $\frac{a}{b} > \frac{d}{e}$

$\frac{dx}{dt} = x(3 - x - y), \quad \frac{dy}{dt} = y(2 - x - y)$

(ii) $\frac{a}{c} > \frac{d}{f}$ and $\frac{d}{e} > \frac{a}{b}$

$\frac{dx}{dt} = x(3 - 2x - y), \quad \frac{dy}{dt} = y(2 - x - y)$

(iii) $\frac{d}{f} > \frac{a}{c}$ and $\frac{d}{e} > \frac{a}{b}$

$\frac{dx}{dt} = x(2 - x - y), \quad \frac{dy}{dt} = y(3 - 2x - y)$

(iv) $\frac{d}{f} > \frac{a}{c}$ and $\frac{a}{b} > \frac{d}{e}$

$\frac{dx}{dt} = x(2 - x - y), \quad \frac{dy}{dt} = y(3 - x - y)$

Equilibria occur whenever *both* $\frac{dx}{dt} = 0$ and $\frac{dy}{dt} = 0$, i.e., whenever a ***v*** nullcline intersects an ***h*** nullcline.

The equilibria are labeled on these phase portraits of Figure 8, and the darkest curves are separatrices of saddles (these are indeed trajectories as well, but exceptional ones, that divide behaviors).

What is interesting about the possibilities shown in Figure 8 is that in three of the four cases, one or the other species may become extinct. Which is more desirable? Well, that depends on whether you are trying for pest control to eradicate population *y* or are trying to save species *y* from an invasive species *x*. We see that it is possible for both species to coexist only in case (ii), and that in case (iii) *either* species may die out, depending simply on the initial populations.

Note that some of the parameters *a, b, c, d, e,* and *f* may be controllable in a real world scenario. Perhaps one can change a food supply, or fence a larger or smaller area to affect crowding. Students can try out various scenarios and suggest a plan to achieve the desired goal, and be able to see the effects immediately.

Competition from Other Viewpoints

An interesting question to pose to students is to consider the competition model in other contexts. For instance, what does this model say about competition between two companies competing for market share, or two countries competing for resources?

In only one of the four scenarios, Figure 8ii, is coexistence of both parties possible! In all three of the others, one entity is extinguished. Moreover, in Figure 8iii, *both* entities are at risk – it would appear that one or the other party is forced into bankruptcy, depending not on its value, but on the initial condition!

Perhaps competition in the marketplace is not as desirable goal as we are led to believe.

CHAOTIC BEHAVIORS

Nonlinearity causes increasingly complicated results. However, three dimensions are necessary for these dramatic effects to occur.

This can occur in a two-dimensional system with time dependence, as we shall show with the forced, damped pendulum, For this example there exist regions of initial conditions with great instability, and we will see that chaotic behavior can occur.

In three-dimensional nonlinear systems, strange attractors can also occur, which we shall demonstrate briefly in a later section.

It is nonlinearity that produces these bizarre effects, but linearization is of scant help for these systems. Rather, it is dynamic software that lets us *see* and tackle chaotic behavior, sensitivity to initial conditions, and strange attractors. And it is the dynamic interactive graphics that makes their study available to students, even at the beginning undergraduate level. We shall first look at pendulum examples.

The Pendulum: Phase Portraits and Solution Graphs

Our favorite dynamic software for starting a study of pendulum motion is again *IDE* We see the simultaneous evolution of the phase portrait, the time series graphs for position, velocity, and acceleration, as well as a simulation of the pendulum motion, as shown in Figures 9 (for the simplest version, of the simple harmonic oscillator—no chaos here—autonomous system) and in Figures 10 and 11 (for the most complicated model, the forced and damped pendulum – non autonomous system, where chaos rears its "ugly" head).

The equations of motion of various types of pendulum motion are given as 2nd order DEs, which we enter as a system of first order differential equations:

$$\frac{d\theta}{dt} = \dot{\theta}, \frac{d\dot{\theta}}{dt} = \ddot{\theta}.$$

Figure 9. Simultaneous evolution of various views of a pendulum under a simple harmonic oscillator (SHO) model

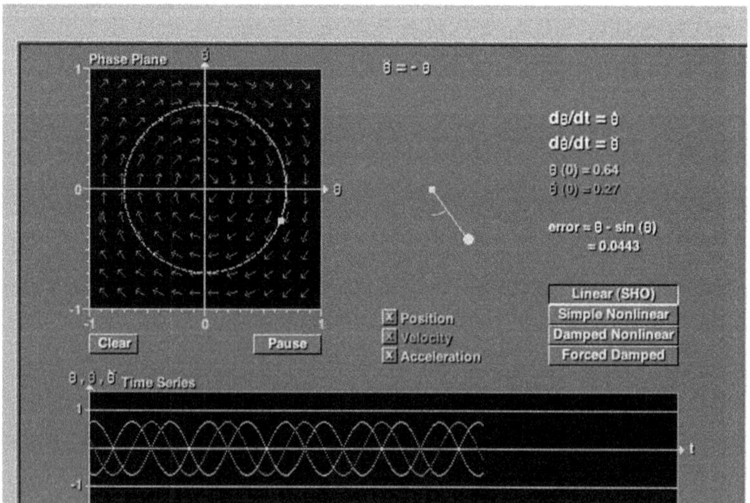

All the user has to do is to choose from the buttons on the right which variation—linear or nonlinear (undamped, damped, or forced)—to watch, then click an initial condition on the phase plane. The motions evolve, slowly enough to *see* the evolution, but quickly enough that it is trivial to replay as many times as desired, in order to get a complete intuitive feel for the various views of the motion and the relation between them. You can watch the graphs evolve at

http://www.aw-bc.com/ide/idefiles/media/JavaTools/pndulums.html

This dynamic exploration is even more important in the case of the forced and damped pendulum, when *t* enters the right side of the differential equation:

$$\theta'' = -\sin(\theta) - b\theta' + A\cos(\omega t)\sin(\theta).$$

The final forcing term includes $\sin(\theta)$ adjusting for the fact that the simulated forcing is in the *vertical* rather than angular direction. See Figure 10.

Figure 10. An exploration of the forced, damped pendulum. The forcing causes the pivot point of the pendulum to move up and down, and the motion is chaotic, often swinging the pendulum over the top.

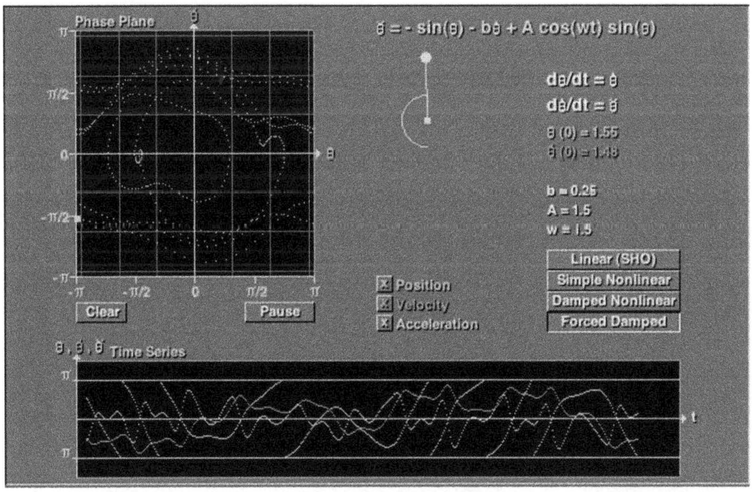

This orbit is chaotic. The phase plane is set for -π to π, which would normally be a period for the pendulum angle theta. But this trajectory is *not* periodic, because of the forcing term $A\cos(\omega t)\sin(\theta)$. The *Pendulums* program "wraps" the orbit – that is, the trajectory continues until it reaches the right side, then it comes back on the left to keep going.

In fact, the forced damped pendulum is *extremely* chaotic. After this orbit ran overnight, we found the screen shown in Figure 11. We see that the tracks all fall within the same odd shaped area, and they have crossed nearly every pixel at this time. Yet the time series also continues to evolve in a fashion similar to that of Figure 10 (The time series graph also wraps around, but when the curves reach the right end, they erase before continuing from the left.).

This interactive sort of illustration is ideal for understanding pendulum behaviors in general, and students profit greatly from the comparisons of these four models.

We see now that pictures can become very complicated, but that does not mean that we cannot understand them, or that they are not important. We can see from the various pendulum models that some solutions are periodic, and some are not. The periodic solutions are among the stable ones that we often seek. The forced damped pendulum is an interesting case in that the solutions do not go off to infinity, but they are chaotic and unpredictable within a bounded region.

Figure 11. The forced, damped pendulum after running overnight

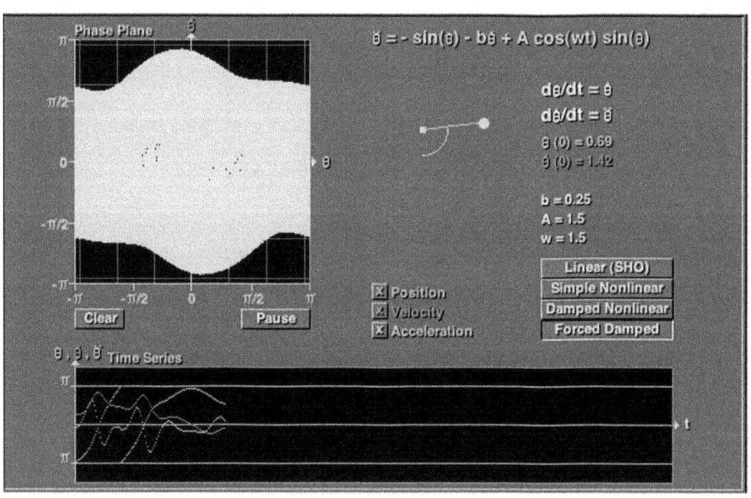

Stability

The behavior of the forced damped pendulum leads us to the central issue in nonlinear DEs – that of *stability.* (Hubbard, 1999, 2008) Stability considerations are essential for such diverse fields as robotics, airplane design, cardiology, earthquake resistance, and sending space ships to the outer reaches of our solar system.

A system is *stable* if small perturbations yield the same long-term behavior. However, contrary to intuition, it is *instability* that is a *necessary* ingredient for many real world situations. Instability is required to attain efficient, powerful, flexible, and maneuverable solutions to varying initial conditions.

As an example, consider learning to ski. The novice, quite fearful, plants skis far apart and tries to maintain a rigid posture, as shown in Figure 12a. But as soon as this skier starts to move on a slope or a curve, he/she loses control and falls down.

In contrast, observe the competitive slalom skier, Figure 12b, who weaves almost effortlessly around all the obstacles and finishes with very little exhaustion. The key is putting the skis close together and maintaining a very flexible posture. In this manner the skier keeps right on the edges of instability but can, with only slight effort, react in order to keep "upright" and continue on to the next obstacle.

A similar analogy applies to cardiology, where the heart must constantly react to changing initial conditions and somehow keep pumping.

A healthy heart does not beat in a perfectly regular pattern; in fact, such a pattern would be a sign of potentially serious pathologies. The body's autonomic nervous system uses neurotransmitters to speed up or slow down the heart, and tiny fluctuations in those substances induce variability in the intervals between consecutive beats (Cain, 2011).

Hurricane prediction profits from recognizing the chaos and sensitivity to initial conditions that are inherent in weather patterns. See the explanation of Ensemble Models (1995) provided in the United States by the National Oceanic and Atmo-

Figure 12. Stable (a) and unstable(b) approaches to skiing

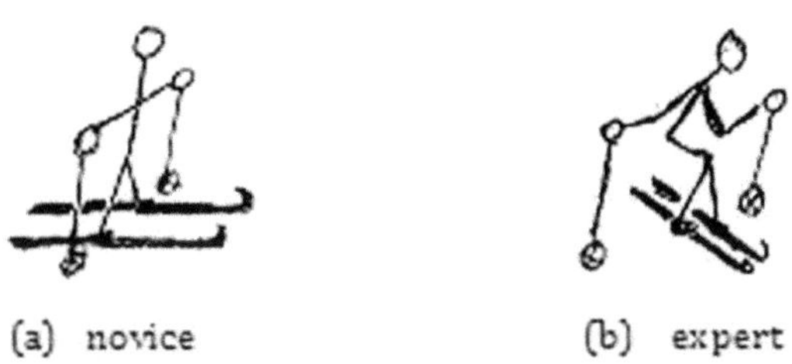

spheric Administration (NOAA) through NCEP (National Centers for Environmental Prediction).

We do not have space here to go into further detail, but we want to be sure that the reader grasps the importance of learning to understand stability and instability. These mechanisms are not a "frill" – it is ever more necessary to handle subjects like these. These examples have been given to show how important it is that DE courses include nonlinear and chaotic behaviors.

The software we have used, as illustrated in the forced pendulum example and the next example (in higher dimensions) is basic to grasping key principles of these behaviors; it is so simple to use that students are easily propelled to a level previous generations met only in graduate school. Furthermore, *MacMath* has been sought and used by a number of nonacademic agencies, such as the Federal Drug Administration (FDA), National Institutes of Health (NIH), United States Coast and Geodetic Survey (USCGS), and many others.

To drive home these points, we take a quick look at higher dimensional systems.

HIGHER DIMENSIONAL SYSTEMS

We have seen that moving from linear systems to nonlinearity even in two-dimensional autonomous systems creates complications of multiple equilibria and periodic limit cycles. Adding another dimension allows for chaos, period doubling, and strange attractors.

We have already seen the chaos created in the forced damped pendulum by incorporating the time variable explicitly in the 2D system, creating a three-dimensional non-autonomous *xyt* system.

An autonomous 3D system of three time dependent variables *x, y,* and *z* also provides new complications. Most of us have heard the story of the Lorenz attractor and its discovery by sensitivity to initial conditions. Some details are given in Hubbard & West (1995) or Farlow et al (2007); excellent further exploration is provided in the lab workbook provided within Hohn et al (2000).

Here we look at a simpler example of a strange attractor, by Sprott (1994) and stated in this form by Linz (1997). This 3D system comes from a third order differential equation, where the x''' term is called a "jerk" (Gottlieb, 1996) because it describes the rate of change of the acceleration, x''.

$$x''' = -x'' + xx' - ax + b$$

or

$$\frac{dx}{dt} = y$$
$$\frac{dy}{dt} = z$$
$$\frac{dz}{dt} = -z + xy - ax - b$$

We rely on graphics to understand this system, and we quickly found an interesting change of behaviors. We show a selected few results of our exploration for $b = 1$ and four values of $a = 0.9, 1.0, 1.2$, and 1.3. Figures 13 and 14 give a picture of typical trajectories in the 3D *xyz*-system, each starting from a point click in the *xy*-plane at approximately (0, -2, 1.8).

Needless to say there are dozens of avenues for further exploration. The sample time series in Figure 15 enhance the understanding of Figures 13 and 14.

Figure 13. Three-dimensional views, n xyz-space. The first frame shows a trajectory spiraling slowly into a sink. The second frame shows the same, except that the spiral evolves ever so more slowly. The third frame shows a limit cycle has appeared, and in the fourth frame, we see "period doubling", where the cycle makes a big loop crossing a small loop (In the fourth frame, the outer loop goes outside the 3D region shown, but we can clearly see it closed in the final xy-phase portrait of Figure 14.).

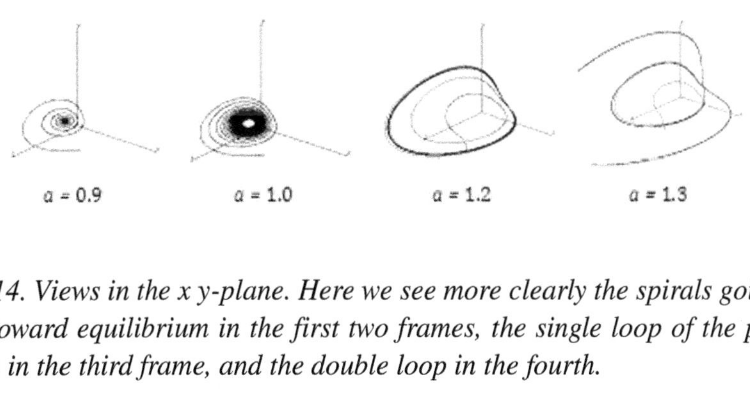

Figure 14. Views in the x y-plane. Here we see more clearly the spirals going very slowly toward equilibrium in the first two frames, the single loop of the periodic solution in the third frame, and the double loop in the fourth.

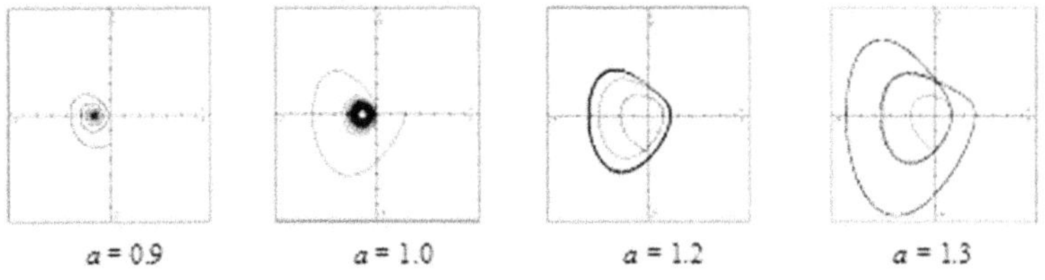

Figure 15. Time series for y(t). For the second frame, the spiraling toward the equilibrium is so slow that we increased the time axis from 50 to 500, to try to show that the amplitude is indeed continually diminishing. You may have to do this experiment yourself to be convinced!

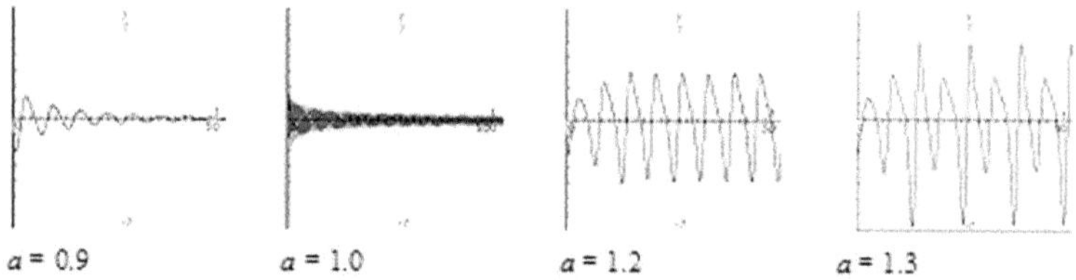

Figure 16. Rotations for the 3D xyz graph when a = 1.3: four selected views

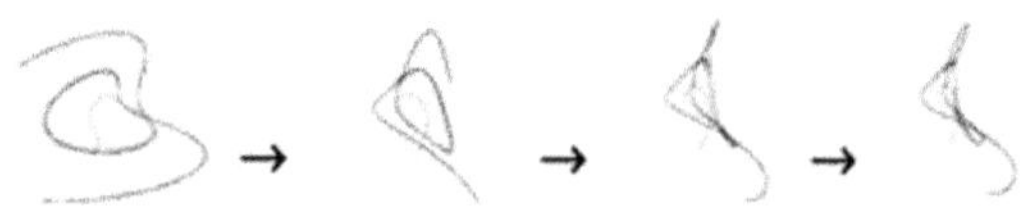

Figure 17. Two trajectories for 3D system, one from outside, one from inside the limit cycle. The time series are graphed together to show that although they settle into the same amplitude, they are offset in the t direction by the different values of y(0).

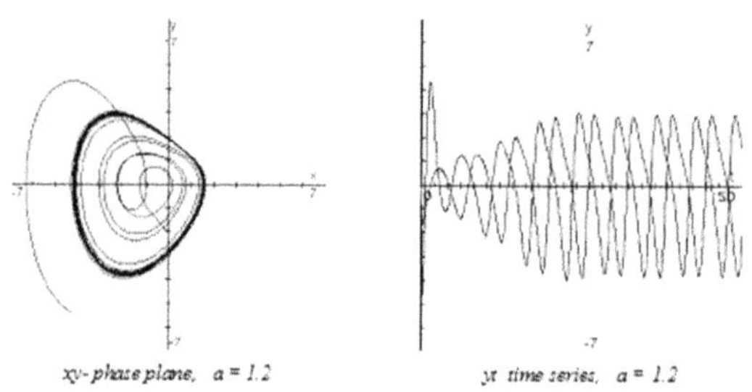

MacMath (Hubbard & West, 1995, 1999) will also rotate the three-dimensional views. Figure 16 shows a sampling of the views you will see if you rotate the fourth frame of Figure 13. Although the outer loop remains broken, you can get an idea of how the figure is warped in three dimensions.

Finally, in Figure 17, we look more closely at the third frame of Figure 14, to convince you that the limit cycle is approached from both the outside and the inside. Each of these frames shows a single trajectory.

We hope you are convinced that these explorations are worthwhile, simple to obtain, and empower the students mightily. They are just a very small sample of the opportunities that await, but we hope they are enough to intrigue you to try for yourself.

SOLUTIONS AND RECOMMENDATIONS

Get interactive graphic software into *every* DE course!

Don't stop your course short of nonlinear DE systems. Give students the basic tools—linearization and graphics—and experience with using them. These are easy extensions of the standard material, but students need enough experience to see how easy it is and what to look for to use these tools intelligently. With visualization tools in hand, you can save a great deal of time on other parts of the syllabus.[3]

Encourage—no, insist on—*discussion* of solution *behaviors*. Realize that working to understand this aspect of DEs, particularly *nonlinear* behaviors, is necessary and that graphics are essential to this effort. Furthermore, this understanding becomes accessible to students if software is dynamic, allowing interaction and exploration.

Focus on long term *behaviors*. One cannot assume that linearizations are valid everywhere near an equilibrium – the long term trajectories can be very different.

Interactive illustrations such as IDE, or those by MIT, or by Blanchard, Devaney and Hall (2006) pay big dividends when incorporated into DE courses. See Miller and Upton (2008).

An important mode of exploration is to see the effects of changing parameters. Real time change of parameters is by far the best way to understand the results of such changes. Static pictures are hard to compare, especially if they can be seen only one at a time.

Guided explorations such as *IDE* (Hohn, et al., 2000) are designed to concentrate on single concepts, maximizing ease of operation while minimizing distraction. Deliberate choices have been made not to include too many options. The mantra for creating interactive illustrations is to "keep it simple." Though you may argue that the screens of Figures 9. 10, and 11 do not look all that simple, they purposely include all the ingredients that must be coordinated just for a basic understanding.

However, we also see what a guided illustration does *not* do – you cannot enter your own equations, nor determine initial conditions very precisely, nor can you necessarily vary the values of all parameters. Thus although IDE is valuable for a first course, one also needs an open-ended software package like *MacMath* (Hubbard & West, 1992, 1999) to explore any model in greater detail, as we did for Figures 5b, 6, and 7, and then with the competition models of Figure 8, and the chaotic models of Figures 14-17. Other open-ended packages with similar capabilities include, in order of origin, *ODE Toolkit* by Harvey Mudd College (2011), *Dfield* and *Pplane* by John Polking, (2005) *ODE Architect* by C-ODE-E (1998), and *Ode Factory* by Jerry Tutsch (2011).

My favorite sources of ODE systems to explore are those by Tony Danby (1985, 1997). His books focus on behaviors, and experiments, as well as offering the basics of differential equations and computer programming for them.

Challenges for Open-Ended ODE Software

Open-ended graphic software that is dynamic and interactive (rather than simply creating static pictures) is of paramount importance for experimentation and research. That which is available today at the introductory student level has yet to match the capabilities of *MacMath*, which has become outdated by advances in personal computer operating systems. Key elements for successful experimentation begin with the following:

- Ability to *enter one's own DE,* using standard mathematical notation.
- *Interactive graphs of phase plane and solution graphs,* where trajectories and solutions for the DE evolve from *a click on an initial condition*. (an alternative numerical mode for entering initial conditions is desirable as well, but is too clumsy to have as the only method – it simply takes too long to enter, especially for multiple points, which distracts the focus from seeing the instant evolution of *behaviors*).
- *Choice of numerical method*, and *control of stepsize*.

Additional options increase the usefulness in classroom or publishing settings:

- *Choices* in Drawing Trajectories, such as "forward", "backward", or "both; options to stop a trajectory at an equilibrium, at a boundary, or after *n* steps.
- *Editing ability for a trajectory or solution* – changing its thickness or color can be immensely helpful for illustrating certain features.
- Ability to *add parametric curves* (Such as isoclines, analytic solutions, or some other curve for comparison with solutions and trajectories).
- Ability to *change parameters* – with sliders, clicks on a parameter plane, or entering values numerically – to see immediately their effect on the graphs. This is how we locate precisely where bifurcations take place.
- Ability to *"Flow"* an entire locus of initial conditions, such as a curve, a circle, a disk, or another shape.
- Ability to *print*, for homework and writing; to export high quality pictures for publishing.

Many existing DE software packages include some of the above, but usually not all. However, what is distressing is that so many such packages do *not* include the following items that we find essential. Polking (2005) and Tutsch (2011) do in fact have these capabilities, although in a somewhat clumsy "add-on"fashion.

- **Precise Location of Equilibria:** We click on a point near where we suspect an equilibrium, and *MacMath* uses Newton's method[4] to locate exact coordinates.
- **Separatrices for a Saddle:** Draw automatically, using initial conditions very close to the equilibrium, in the direction of the eigenvectors. These are necessary to show precisely the basins of attraction of sinks or limit cycles. Only then one can see exactly where a linearization is a reasonable approximation and where it is not.
- **Linearization Information:** Essential data can be provided automatically: coordinates of the Jacobian matrix, its eigenvalues, and eigenvectors. All of this information, including, the type of equilibrium, can be displayed in a popup window.

These are the features that give the real power to students of differential equations. Our fervent hope is that these features will become universally available very soon – only then will we be able to feel our quarter century of effort has paid off.

A more difficult question is how to convey information in higher dimensions. Just in three dimensions, you might imagine how a boiling pot of spaghetti tangles the noodles up in all directions – differential equations take full advantage of this opportunity.

We have shown some of the *MacMath* features for a 3D system in Figures 14-17, including rotations of a three dimensional representation. We can also show Poincaré sections, and study maps that iterate the numerical approximations.

Possibly for dimensions above three, systems are inherently too complicated to hope to understand them completely. However, for a system with *any* number of first order equations, the programming of numerical approximations is the same, with simply more columns in the "spreadsheet" behind the scenes of the graphs. As with *Microsoft Excel,* it should then be a simple matter to create a two dimensional graph of any combination desired. Just being able to *see* graphs of $x(t)$, $y(t)$, $z(t)$, $w(t)$, ... can give a great deal of information, especially for a system without formula solutions.

FINAL REMARKS

The revolution that computer graphics brought to the study of differential equations occurred over a quarter century ago. Yet somehow even today many ODE courses are taught without computer graphics exploration, relying only on formulas that don't even solve the majority of differential equations.

We converted *many* colleagues, but somehow the effort stopped spreading, and many students in the current generation are denied the understanding and power that comes from learning to interact with the pictures.

The message that needs to get through is that it is the behaviors, not the formulas, that matter.

The good news is that indeed linearization is key to understanding nonlinear DEs, but that is *not* enough. We *need* the overall pictures to see just how the linearizations fit together at multiple equilibria, to know exactly where a linearization is valid. We also need pictures to explore the more complicated effects -- chaos, strange attractors, bifurcations, become clear only with the aid of visual representations.

REFERENCES

Blanchard, P., Devaney, R. L., & Hall, G. R. (2011). *Differential equations* (4th ed.). London: Brooks Cole.

Borrelli, R., & Coleman, C. (2004). *Differential equations: A modeling perspective*. New York: Wiley.

C-ODE-E (NSF Consortium for Ordinary Differential Equations Experiments). (1998). *ODE architect*. New York: Wiley.

C-ODE-E (NSF Consortium for Ordinary Differential Equations Experiments). (2011). *Website and newsletters (from 1994)*. Retrieved from http://www.codee.org

Cain, J. W. (2011). Taking math to heart: Mathematical challenges in cardiac electrophysiology. *Notices of the American Mathematical Society*, *58*(4), 542–549.

Danby, J. M. A. (1985). *Computing applications to differential equations: Modelling in the physical and social sciences*. Reston, VA: Reston Publishing Co.

Danby, J. M. A. (1997). *Computer modeling: From sports to spaceflight... from order to chaos*. Richmond, VA: Willman Bell, Inc.

Farlow, J., Hall, J. E., McDill, J. M., & West, B. H. (2007). *Differential equations and linear algebra* (2nd ed.). New York: Pearson.

Gottlieb, H. P. W. (1996). What is the simplest jerk function that gives chaos? *American Journal of Physics*, *64*, 525. doi:10.1119/1.18276.

Harvey Mudd College. (2011). *ODE toolkit*. Retrieved from http://odetoolkit.hmc.edu/

Hohn, H., West, B. H., Strogatz, S. H., McDill, J. M., & Cantwell, J. (2000). *Interactive differential equations*. Reading, MA: Addison Wesley. Retrieved from http://www.aw-bc.com/ide

Hubbard, J. H., & Lundell, B. E. (2011, March). A first look at differential algebra. *The American Mathematical Monthly*, 245–261.

Hubbard, J. H., & West, B. H. (1991). *Differential equations: A dynamical systems approach, part I: The one-dimensional theory*. New York: Springer Verlag.

Hubbard, J. H., & West, B. H. (1992). *MacMath 9.2*. New York: Springer Verlag.[5]

Hubbard, J. H., & West, B. H. (1995). *Differential equations: A dynamical systems approach, part II: Higher dimensional systems*. New York: Springer Verlag.

Hubbard, J.H., & West, B.H. (1999). *MacMath 10.0*. Unpublished.[5]

Linz, S. J. (1997). Nonlinear dynamical models and jerky motion. *American Journal of Physics*, *65*, 523–525. doi:10.1119/1.18594.

Miller, H. R., & Upton, D. S. (2008). *Computer manipulatives in an ordinary differential equations course: Development, implementation, and assessment*. Retrieved from http://math.mit.edu/mathlets/wp-content/uploads/cet-published.pdf

MIT Mathlets. (n.d.). Retrieved from http://math.mit.edu/mathlets/

MIT OCW Differential Equations Course. (2010). Retrieved from http://ocw.mit.edu/courses/mathematics/18-03-differential-equations-spring-2010/

MIT Open Courseware. (n.d.). Retrieved from http://ocw.mit.edu

NOAA. (1995). Ensemble forecasting at NCEP. *National Centers for Environmental Prediction, National Oceanic and Atmospheric Administration*. Retrieved from http://www.emc.ncep.noaa.gov/gmb/ens/info/ens_detbak.html

Polking, J. C. (2005). *Dfield, and pplane*. Retrieved from http://math.rice.edu/~dfield/dfpp.html

Sprott, J. C. (1994). Some simple chaotic flows. *Physical Review E: Statistical Physics, Plasmas, Fluids, and Related Interdisciplinary Topics*, *50*, R647–R650.

Tutsch, J. H. (2011). *Ode factory*. Retrieved from http://odefactory.com/

ADDITIONAL READING

West, B. H. (Ed.). (1994). Special issue on teaching differential equations. *The College Mathematics Journal, 25*(16).

West, B. H. (1999). Technology in differential equations courses: my experiences, student reactions. *MAA Notes, 50*, 78-79. Washington, DC: Mathematics Association of America.

ENDNOTES

1 For $dx/dt = f(x,y,t),\ dy/dt = g(x,y,t)$, if *f* and *g* are continuously differentiable (or, more broadly, functions that satisfy a Lipschitz condition), then through any point (*x,y,t*) there exists one, and only one, solution to the DE (Hubbard & West, 1995).

In an autonomous system of DEs, $dx/dt = f(x,y),\ dy/dt = g(x,y)$, phase plane trajectories will not cross. In a *nonautonomous* system, where the DE has explicit time dependence in *f* and/or *g*, phase plane trajectories in the *xy*-plane *can* cross, but the uniqueness/existence theorem assures us that only one trajectory goes through any point in *xyt*- space. Software that will plot and rotate trajectories in 3D allows you to *confirm* this fact.

2 The Euler method is the most straightforward numerical method, a simple linear approximation that can go astray where the vector field curves. Variations on the Euler method, or Runge Kutta methods, are fairly simple averaging methods that improve accuracy. A myriad of other possibilities include multistep methods and other algorithms.

3 Examples of saving course time are using interactive illustrations such as the mass and spring series, the pendulum series, or the linear classification (parameter plane) series as assignments *before* the usual textbook material – this makes these topics go a *lot* faster. Another time-saver is not to do *every* analytic solution method – separation of variables and solving linear DEs are the most useful techniques.

4 Of course Newton's method does not always converge to the closest equilibrium, but students can understand this and, using the pictures as a guide, can easily distinguish the cases where the calculation converges elsewhere.

5 MacMath 9.2 and MacMath 10.0, alas, only run on the Mac Classic operating system. But the programs for 2D differential equations and 2D iterated maps are available at http://www.math.cornell.edu/~dynamics/BenHinkle/index.html

Chapter 6
Vectors and Differential Equations:
A Visual Approach using Autograph

Douglas Butler
iCT Training Centre, UK

ABSTRACT

This chapter is inspired by a session the author gave at the 10th EMAC Conference (Engineering, Mathematics, and Applications) at the University of Technology, Sydney, in December 2011. The audience was university teachers, but the software, Autograph, was designed for use in High Schools. The author was able to show how a simple, pedagogically focused interface could be used to create a highly visual approach to the teaching of two favourite topics: Vectors (in 2D and 3D) and Differential Equations (1st and 2nd Order).

INTRODUCTION

Teachers of University mathematics are generally used to using one of the software giants, such as Maple, Mathcad, or Mathematica. These tools can be tamed to perform almost any mathematical task, but each requires a detailed knowledge of syntax; furthermore, the presentation is generally in "notebook" style, which is perfect for research use, but can appear confusing when used as visualizer in the

DOI: 10.4018/978-1-4666-4050-4.ch006

Figure 1. Autograph has a cleanly designed menu/icon system for its statistics, 2D and 3D pages

classroom. Autograph on the other hand is designed with teaching in mind: the screen is uncluttered, and all operations are easily found by the judicious use of "select" and "right-click".

The University teachers in Sydney were somewhat surprised that software created for high school use could bring so many new insights into the teaching of vectors and differential equations (See Figure 1).

BACKGROUND

Autograph was conceived in the secondary classroom, covering topics right up to University entrance in the UK. This included many topics that have long since risen above school curriculums in the general rush to consolidate, together with the laudable aim of trying to encourage more students to study mathematics. The two topics of this chapter, vectors and differential equations, used to be well inside the school envelope, but they now occupy the overlapping space between the top end of high school and the first year of university mathematics.

In the design of Autograph, there was an overriding determination to make the creation of dynamic objects as straightforward as possible. This has led to an environment where it is possible to study concepts from the "bottom up". The logical addition of dependent objects and variable parameters can help students to build firm foundations based on a creative mix of dynamic visualisations and solid bookwork.

The get the most out of this article, you are invited to download and install Autograph (PC or Mac) from www.autograph-maths.com/download and to email the author for a serial number (otherwise your installation will run out after 30 days).

Please also refer to the online dynamic full colour version of this chapter on the Web: www.tsm-resources.com/dynamic.

Figure 2. A clear understanding of elementary operations can be gained if the vectors move

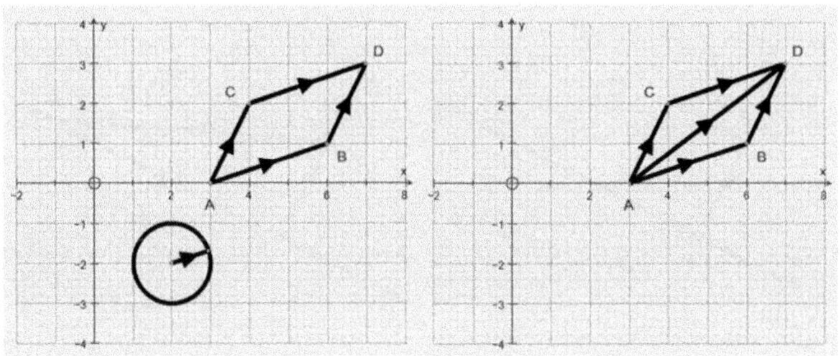

Representing Vectors in 2D

I am a firm believer in teaching vectors in 2D the way they are handled in 3D. This way the transition from 2 to 3 dimensions can be painless.

Figure 2 illustrates a strategy for understanding the principles of vectors from a standing start. In Autograph you can define a vector on a given point by entering its two components, e.g. define AB = [3,1], and the point "A" is moveable, or define the vector between two points, e.g. A (3,0) and B (4,2), in which case both A or B are moveable.

Moving the point A shows that the vector [3, 1] can in fact be anywhere, including copied onto C (to form CD). Likewise, AC can be copied to BD. Hence, the parallelogram law can be studied, but only once we have established addition. The addition of two vectors is defined as a vector with the sum of the x and the sum of the y components, and hence can be represented by the leading diagonal of the parallelogram.

To move on to subtraction, we recall that in simple addition of numbers, 3 + (–3) = 0: so the process of subtraction requires the concepts of a negative number and of zero to be in place. So it is with vectors: to establish subtraction, we need a "negative" vector and "zero" vector to be defined. Students can be asked to move C until the sum is zero. This leads to the definition of a negative vector and a zero vector, and ultimately to the realization that subtraction can be represented by the other diagonal of the parallelogram.

The concept of a unit vector can also be explored: draw a unit circle on a point somewhere, and create a unit vector on that point from the vector AB (or CD). Note that it is in the same direction, but of unit length.

Figure 3. A dynamic representation of the vector equation of a straight line

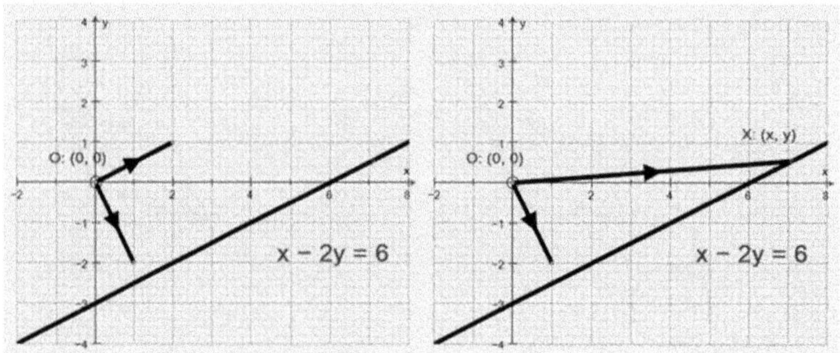

There are many forms for the straight line: explicit (e.g. y = 3x – 1) and implicit (e.g. x – 2y = 6), and based on the vector form: [x, y] = [a, b] + k[l, m], which is a line through (a, b) parallel to the vector [l, m].

It is useful to use the implicit form to prepare students for the equation of a plane, and to establish that the vector [a, b] is perpendicular to the line ax + by = c (as illustrated in Figure 3 on the left). With this straight line, the parallel vector is [2, 1], and so the perpendicular vector is [1, –2]. Is it a coincidence that "1" and "–2" are the coefficients of "x" and "y" in the straight line equation?

To confirm this, place a point "X" on the line. The vector from the origin to X is the vector [x, y]. It is useful to be able to confirm that the scalar product of [1, –2] and [x, y] is 6. More generally, for the line ax + by = c, the value of [a, b].[x. y] is c and [a, b] is perpendicular to the line (as illustrated in Figure 3 on the right).

The Transition to 3D

The transition to 3D should now be a simple matter of adding another dimension. A line becomes a plane and the perpendicular vector now takes its components from the three coefficients of "x", "y" and "z" (Figure 4, left).

Figure 4, right, shows how three points in a plane can create two vectors whose sum also lies in the plane. Furthermore, moving the original vectors around the plane can demonstrate how any point in the plane can be found as a linear combination of the original two vectors. Hence, a plane can be defined from a point and two vectors, for example the upper plane in the diagram.

The Shortest Distance between Two Skew Lines

This is a tricky topic to teach, but one that responds beautifully to a dynamic visual approach. In Figure 5, left, a straight line has been formed from two points, and a

Figure 4. The extension from 2D to 3D is a natural step with dynamic figures

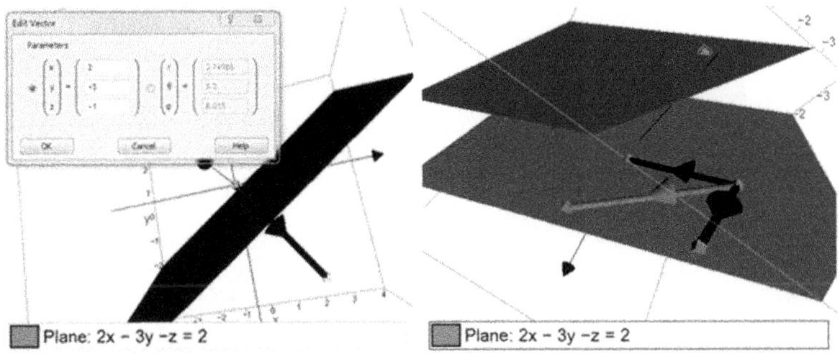

second line entered in vector form, through (1, 2, 0) and parallel to the vector [-1, 1, 1]. The shortest distance is part of a line that is in a direction that is mutually perpendicular to the two lines. To illustrate this, create two vectors, one on each line. Place a point somewhere on one of the lines, and create on this point the cross product of the two vectors.

Now construct a straight line through the end points of this vector. In Figure 5, right, this point has been moved along the line on which it is based until the perpendicular line cuts the second of the two original lines. This new point defines the shortest distance! You can confirm this by getting Autograph to do the calculation: select the two original lines and ask for "shortest distance".

The Line of Intersection of Two Planes

In Figure 6, left, one plane is $2x - 3y = 6$, and the second is defined by a point (1, 1, –1) and the two vectors [1, –1, 1] and [2, 1, 1]. Autograph has been asked to construct the line of intersection, and in doing so states its vector equation as

Figure 5.

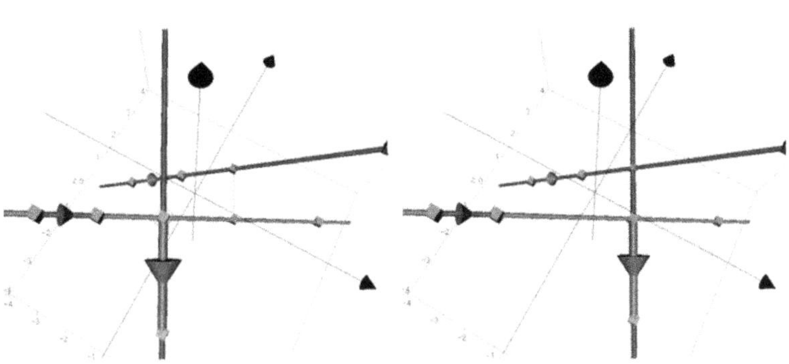

Figure 6.

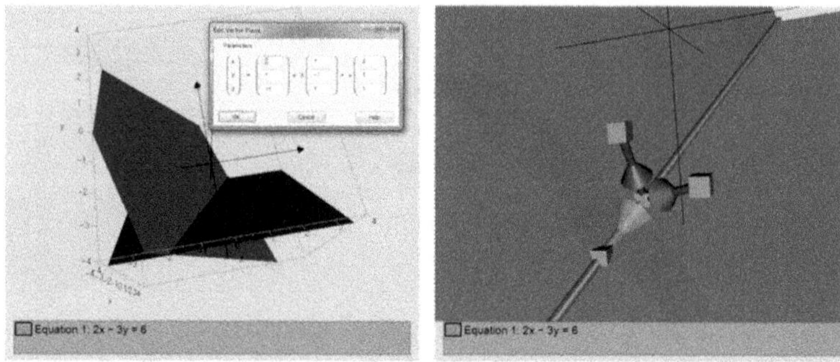

$[x, y, z] = [0, -2, -0.6667] + \lambda[0.7804, 0.5203, 0.3468]$

and the angle between the two planes is 58.74°, or 1.025 radians.

This version of the vector equation is of course only one of an infinite number of possibilities, based on any point on the line, and any vector that is parallel to [0.7804, 0.5203, 0.3468], so this is unfortunately not a good vehicle for checking student answers!

So far so good, but where is the understanding?

If you place a point someone along the line, and then for each plane construct the normal unit vector, you have the opportunity to draw the cross product of these two unit vectors. Autograph gives the answer for this cross product as [−0.6671, −0.4447, −0.2965]. Is this vector parallel to the vector given as part of the intersection line equation: [0.7804, 0.5203, 0.3468]?

This will demonstrate convincingly that the line of intersection is in a direction that is mutually perpendicular to the two planes, as shown in Figure 6, right.

While We Are in 3D ...

Other topics to benefit from 3D graphing include sections of a cone, and various possibilities in the construction of volumes of revolution of an area about an axis.

The teaching of all these topics has been immeasurably improved by software that can treat them in 2D and 3D using dynamic objects.

Visualising Conic Sections

In Figure 7, left, a family of conics has been drawn from the polar equation

$1/r = 1 + k \cos\theta$

Figure 7.

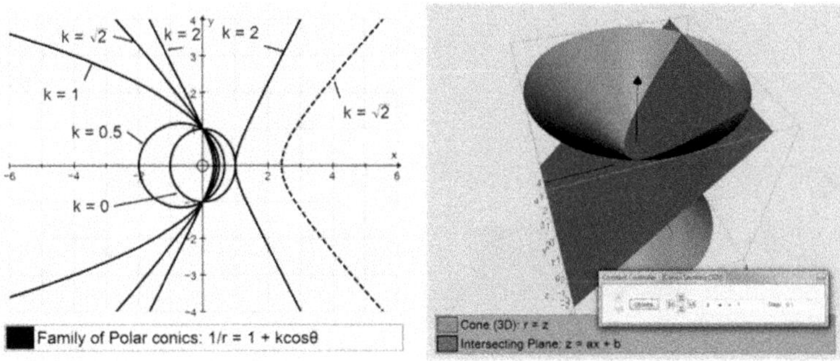

with values of "k" set as follows: k = 0 (circle), k = 0.5 (ellipse), k = 1 (parabola), k = $\sqrt{2}$ (rectangular hyperbola) and k = 2 (hyperbola).

A good knowledge of these important curves in 2D can be achieved through the polar form, though a firm understanding of all the Cartesian forms and their various coordinate geometry properties is also important.

With this grounding in 2D, the 3D scene is a perfect follow-on. Figure 7, right, shows the polar form of a cone, r = z (though the Cartesian form $x^2 + y^2 = z^2$ is another way of plotting the same thing). So how to draw the intersecting plane? In 2D, y = mx + c is a straight line. In 3D the same equation is a plane as it plots y =- mx + c "for all z". The plane that we want is z = ax + b (plotted "for all y"), and varying "a" is just the same as varying "m" as the slope of the line in 2D. Autograph's constant controller can vary "m" to find all the categories of conic section, including of course the double-branched hyperbola.

A New Look at Volumes of Revolution

In Figure 8, a parabola has been drawn (plotted as 2D in 3D), and an element of area created between two points on the curve that are close to each other, which is just a single rectangle. In Figure 8, left, an axis of rotation has been drawn from y = –1 (treated as a 2D object in3D), and in Figure 8, right, the axis is x = –1.

In each case an element of volume is shown, but how important it is to train students to predict these shapes before they are drawn.

Some wonderfully interesting 3D volumes can be created using areas, appropriate axes of revolution, and different "rules" for calculating the area. How important a good grounding in 2D is before tacking the 3D situation. In Figure 9, left, the process of finding the area between two curves is being explored in 2D, and in Figure 9, right, the effect of rotating such an area to form a solid in 3D.

Figure 8.

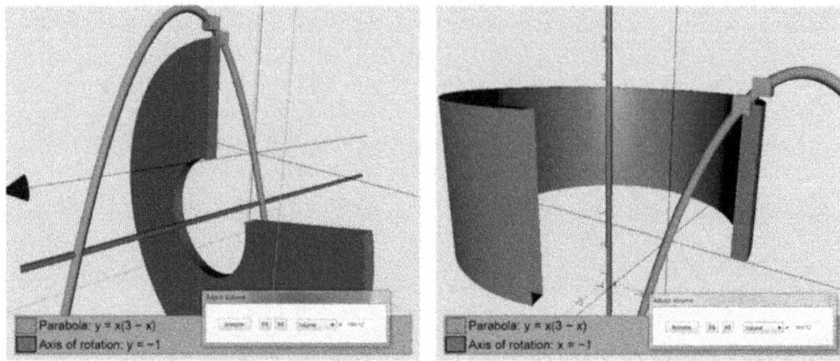

Teaching 1st Order Differential Equations from the Bottom Up!

This has to be a topic that can really benefit from the use of dynamic images. Additionally, there are big advantages for teachers and learners of differential equations to having software available that will instantly plot slope fields from equations that are entered either explicitly or implicitly.

With Autograph, it was decided to make use of the keyboard characters “and ‘to create

$y' = dy/dx$ and $y'' = d^2y/dx^2$

entries. In the case of “t” being the independent variable and “x” the dependent variable, characters x-dot and x-double-dot [$\dot{x}$ and $\ddot{x}$] have been used.

Figure 9. Similarly, the extension of 2D to 2D in calculus is a natural step

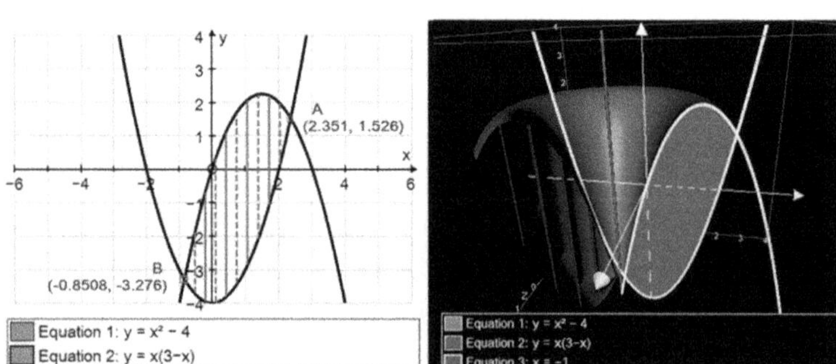

The other interface issue to settle is how to enter the starting conditions. This requires one or more points to act as the starting point for the solution. With Autograph there are three options:

- Start each solution off by a mouse click.
- Use a selected point as the start point, which can then be dragged around.
- Establish and use a set of points on a line which will create a neat family of solutions.

With a tool such as Autograph it is possible to establish a really firm understanding of the basic principles by tackling the subject from the "bottom up". A typical sequence to explore the basic linear first order differential equation is:

1. y' = 0
   The solution set is many horizontal straight lines.
2. y' = 1
   The solution set is many lines straight lines with gradient = 1.
3. y' = x
   This yields is a family of parabolas. You can then compare the slope field with a family of parabolas from $y = \frac{1}{2}x^2 + c$, as in Figure 10, left. In this diagram, a moveable vector [0, 2] has been placed at a point on one of the curves, to show that it is a vertical displacement that defines the family of parabolas.
4. **y' = y**
   Ask what the solution is, and hold your breath as answers such as "parabolas the other way" come in - it may have been only last week you were teaching the exponential function! See Figure 10, right.
   This is of course a nice reminder that $y = e^x$ is a function that is equal to its own derivative, but what are all the other solutions indicated by the slope field? Again, wait for "$e^x + c$" … ! The visual approach confirms that the answer cannot be $y = e^x + c$. So where does the constant of integration go?
5. **y' = –y**
   This is the first time an exponential decay appears. So now we can usefully rearrange implicitly:
6. **y' + y = 0**
   This is also the exponential decay of course, and the basic building block of the whole topic of linear 1st and 2nd order differential equations.
7. y' + y = x
   The left side of this equation is referred to as the "complimentary function"; the right side is the "particular integral". This visual approach confirms that

Figure 10. Introducing 1st order differential equations "from the bottom up"

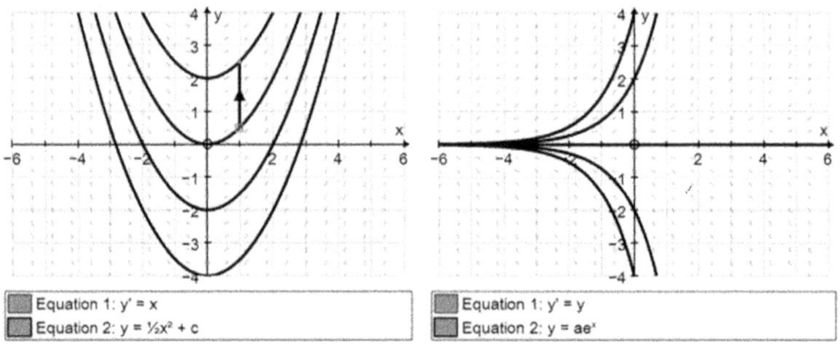

this complimentary function always implies an exponential decay down to the steady state, in this case $y = x - 1$. See Figure 11, left.

8. $y' + y = \sin x$
   This is another good example of exponential decay, this time decaying to a periodic function, see Figure 11, right, which also shows the relationship between the steady state and the particular integral.

Helping students to predict slope fields is as important as it is to ensure that they predict regular graphs before they are computer-drawn. Figure 12 gives two good examples of predictable outcomes.

Other Applications of 1st Order Differential Equations

There are many other applications of 1st order differential equations that can be visualized well; here are two more. First, the plotting of isoclines (points of equal slope), in this case, in Figure 13, left, the family of circles that are related to the differential equation $y' = x^2 + y^2$.

The second example, in Figure 13, right, is the modeling of a falling object that is approaching its terminal velocity. In this case the independent variable is time (t) and the dependent variable is velocity (v). The equation is derived from applying Newton's 2nd Law to the falling mass:

$$mg - kv^2 = m\,dv/dt$$

and the appropriate constant values entered ($m = 80$kg, $g = 9.81$ m/s^2, $k = 2$ and $n = 2$). Again, the ability to enter a differential equation implicitly is a real benefit.

Figure 11. How to visualise the complimentary function and particular integral

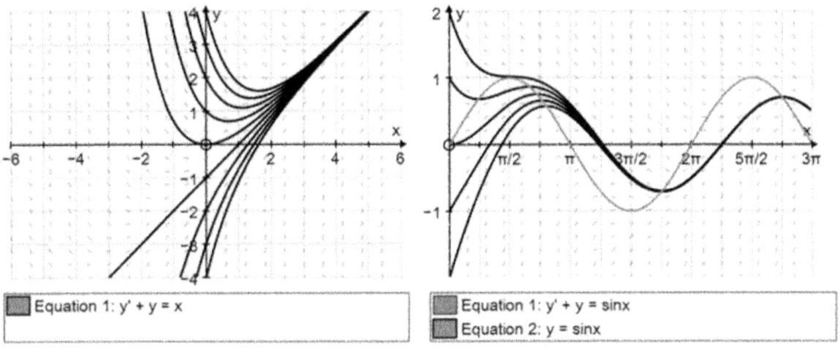

Teaching 2nd Order Differential Equations from the Bottom Up!

If you can establish a firm understanding of graphing and vectors in 2D then the 3rd dimension will follow naturally. Similarly, if you work through 1st order differential equations systematically, and using the implicit form, then the move to second order equations can be painless. Once again, the ability to enter differential equations implicitly is of paramount importance to a firm understanding of their form.

Second Order equations require two starting conditions. Autograph offers the chance to define the first as one or more points at the start, with the same options as the 1st orders (click, selected point or a line of points). The second condition is determined by the initial slope.

This is a typical sequence for "bottom" up learning of the 2nd order equation.

1. Reset the axes to the "x" on the vertical axis and "t" on the horizontal axis, taking account of the fact that the solutions of 2nd orders are usually concerned with displacement and time.

Figure 12. How to predict special instances of elementary 1st order differential equations

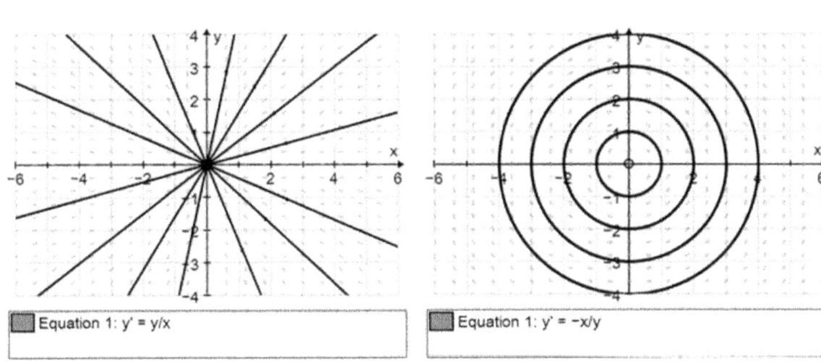

Figure 13.

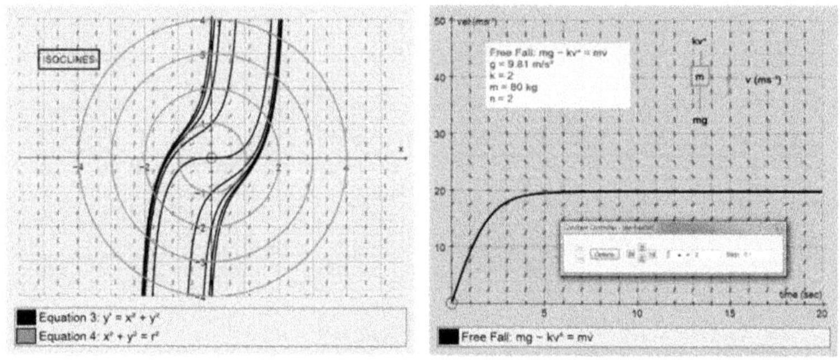

2. $\ddot{x} = 0$
On the first integration, we get $\dot{x} = a$, and on the second integration we get $x = at + b$.
ie The solutions take the form of any straight line through the start point with any initial gradient.
3. $\ddot{x} = 1$
On the first integration, we get $\dot{x} = t + c$, and on the second we get $x = ½t^2 + cx + d$.
ie Any parabola through the start point with any initial gradient.
4. $\ddot{x} = x$ [see Figure 14, left]
The interpretation here is that the acceleration is away from the t-axes whether "x" is positive or negative. The solution take the form $x = a\exp(t) + b\exp(-t)$.
5. $\ddot{x} = -x$. Here the acceleration is always towards the t-axis, and switches as a solution crosses the t-axis, hence there are oscillations which turn out to be sinusoidal. Figure 14, right shows many of these DE forms, all on the same pages, all starting at (0,1) with zero initial gradient.
6. $\ddot{x} + x = 0$
This is a literal rearrangement of $\ddot{x} = -x$, so a sinusoidal solution is expected. In Figure 15, left, the plot of $\dot{x}$ has also been included (as a dotted line).
7. $\ddot{x} + x = f(t)$
This gives the opportunity to explore many different particular integrals
8. $\ddot{x} + 2\lambda\,\dot{x} + n^2x = 0$
This introduces "damping", and the constant controlled is a good way to vary "λ" and hence vary the damping (Figure 15, right).
9. $\ddot{x} + 2\lambda\,\dot{x} + n^2x = f(t)$
This gives the opportunity to explore many different particular integrals.

Figure 14.

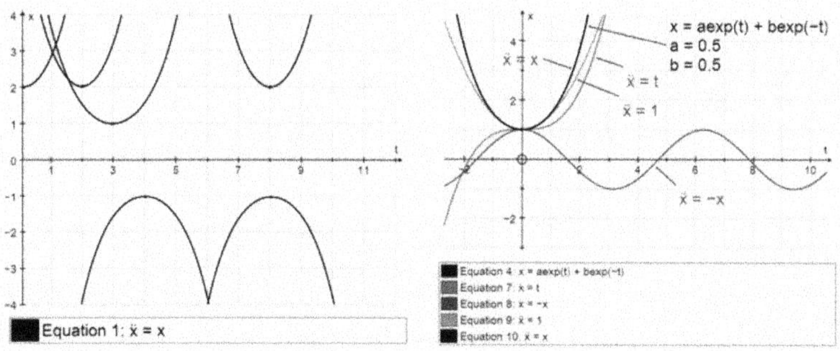

Figure 15. Introducing 2nd order linear differential equations "from the bottom up"

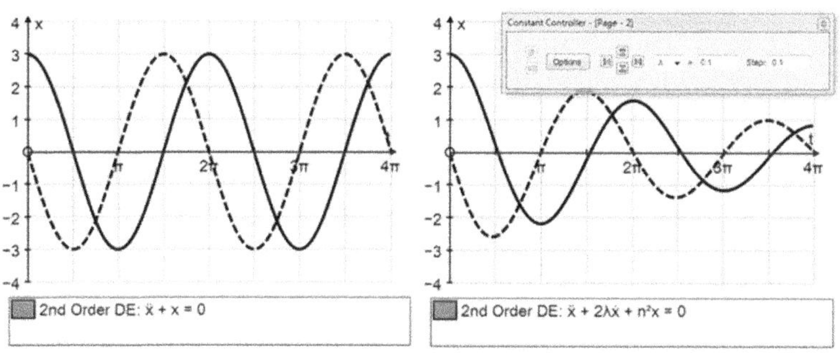

Entering Mathematical Notation, When One-Line Mathematics is Sufficient

There are many occasions when mathematical expressions can be perfectly adequately entered into "Word" as a single line using the standard "Unicode" symbol. The equations for this article were all entered using the Autograph onscreen keyboard (Figure 16).

CONCLUSION

There are many software systems available to take these and many other topics on to a high level, but the approach taken by Autograph is ideally suited to establishing good understanding from the bottom up. Mathematics is, above all, a subject that requires a firm understanding of basic principles. All teachers and students should be wary of any attempt to suggest that modern software and Internet resources can in any way short circuit the learning process.

Figure 16. Keyboard

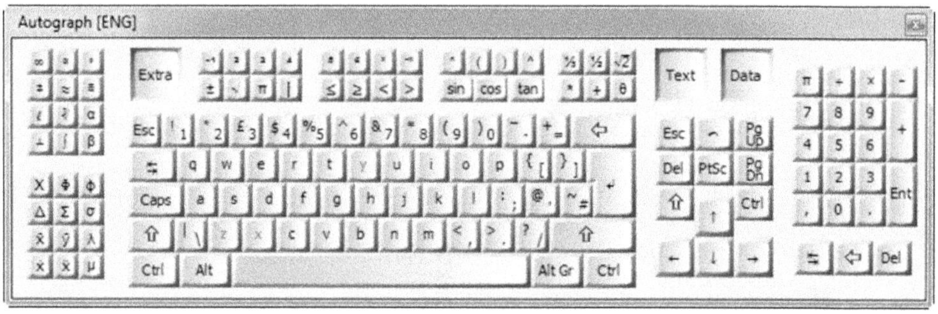

Autograph was designed for the visualisation of school-level and first year university level topics. The two topics in this chapter, vectors and differential equations, and perfect examples of how successful a visual approach can be.

However, a printed chapter is not the ideal environment in which to discuss the impact of dynamic software – its power emanates almost entirely from the fact that things move! There is a dynamic Web page dedicated to illustrating all the examples in the chapter on www.tsm-resources.com/dynamic.

debutler@argonet.co.uk

REFERENCES

Autograph Software. (n.d.). Retrieved from www.autograph-maths.com

Autograph Specific Resources. (n.d.). Retrieved from www.tsm-resources.com/autograph

Butler, D. (n.d.). *Blog*. Retrieved from http://dougbutler.wordpress.com/

Dynamic Online Content of this Chapter. (n.d.). Retrieved from www.tsm-resources.com/dynamic

TSM Resources. (n.d.). Retrieved from www.tsm-resources.com

Chapter 7
Interactive Applets in Calculus and Engineering Courses

Heidi Burgiel
Bridgewater State University, USA

Haynes Miller
MIT, USA

Chad Lieberman
MIT, USA

Karen Willcox
MIT, USA

ABSTRACT

This chapter reports on the use of specialized computer applets ("Mathlets") in two different contexts: on-line instruction in calculus through MIT OpenCourseWare and on-campus laboratory exercises on stability of difference schemes in class. Specific applets are described. Three use cases with varying levels of sophistication (elementary, intermediate, and advanced) are outlined.

INTRODUCTION

In the fall of 2000, a group in the MIT Mathematics Department began work on a suite of computer applets ("MIT Mathlets") designed to support basic MIT mathematics education. For a report on this work and a review of relevant literature see Miller and Upton (2008). The current state of this collection, along with supplementary material, can be found at http://math.mit.edu/mathlets/. The intent of this chapter is to discuss the recent design and application of two new families of MIT Mathets serving very different audiences:

DOI: 10.4018/978-1-4666-4050-4.ch007

- Calculus tools designed for self-learners and delivered over the Internet through MIT OpenCourseWare.
- Tools designed for use in an aerospace engineering classroom workshop to aid in understanding the stability of certain difference schemes.

We will begin by describing the pedagogical factors guiding the creation of this family of applets, the history of their early development, and the background for the recent additions. The next section is devoted to the design and usage of calculus applets for self-learners, and the section that follows to similar treatment of the Eigenvalue Stability applet built for use in an Aerospace Engineering class. We wrap up each section with some lessons learned and a sketch of future work in this direction.

BACKGROUND

Development of the MIT Mathlets began in the Fall of 2000 under a grant from the MIT d'Arbeloff Fund for Excellence in Education. Initially they focused on material for the large course on Ordinary Differential Equations. In the spring of 2003 they were first introduced in that course as lecture demo material and, more powerfully, as a framework for homework assignments. This initial phase of development is reported on by Miller and Upton (2008, available also online), where a detailed study of the pedagogical impact of the applets can be found. As described in that paper, the initial MIT Mathlets began as modifications of applets created for an earlier project, Interactive Differential Equations (2008), authored by Beverly West, Steven Strogatz, Jean Marie McGill, and John Cantwell. The applets for IDE (short of Interactive Differential Equations) were designed and programmed by Hu Hohn, Director of the Computer Arts Center at Massachusetts College of Art, and Hohn continued the development of this initial group and many subsequent applets for the MIT project.

Several well-defined overarching design elements, described in Miller and Upton (2008) shaped the creation of these tools and continue to shape the new ones. They may be summarized as follows. Each applet

- Responds in real time to slider or cursor movements determining system parameters.
- Is narrowly focused, providing maximal insight into important concepts using carefully selected examples.

- Is designed to maximize the visual connections among components, using color coding, layout, and other methods.

In the ensuing years many homework problems incorporating these applets have been written (and many are available at the MIT Mathlets site) and further applets were added to the collection. Some of the new applets were again connected with differential equations, but increasingly attention turned to providing material for use in other courses, especially in courses outside the Mathematics Department. Most of our engineering students will have seen these applets in use in the basic differential equations course and will be familiar with their visual conventions and their philosophy of multiple representation and cursor controls. They are therefore easier for these students to use, and provide an attractive link between courses in different departments and different years of the student's career.

Here is a list of recent transdepartmental projects. They were all programmed by Jean-Michel Claus.

- **2007:** With Peter Dourmashkin of the MIT Physics Department, the applet Series RLC Circuit was designed and implemented as part of the TEAL approach to basic instruction in electromagnetism. This applet has subsequently also been used in the Differential Equations course, providing an important and conspicuous link between courses.
- **2007–2008:** Julie Greenberg of the Harvard-MIT Division of Health Sciences and Technology designed the applet Discrete Fourier Transform. It is used in the HST course "Biomedical Signal and Image Processing".
- **2008–2009:** With John Deyst and Karen Willcox (of the MIT Aero-Astro Department) the applet Nyquist Plot was designed and built. A subsequent collaboration with Franz Hover of the MIT Mechanical Engineering Department resulted in the applet Bode and Nyquist Plots.
- **2010:** With Karen Willcox and Chad Lieberman, the applet Eigenvalue Stability was designed and built, and a worksheet guided students through its use in a class in the course Computational Methods in Aerospace Engineering.
- **2010:** With Heidi Burgiel and Jerry Orloff (of MIT Experimental Study Group), a group of applets illustrating basic calculus concepts was designed and built for use with OCW Scholar.

In this chapter, we shall report in detail on the last two developments. These two differ greatly in their pedagogical context and in their target audiences, and together they present a case study of how a common technology can address diverse pedagogical needs.

CALCULUS MATHLETS FOR SELF-LEARNERS

In the spring of 2011, MIT OpenCourseWare launched OCW Scholar, reorganizing and augmenting the OCW content of a series of foundational undergraduate MIT courses to provide a pathway through the course suitable for use by self-learners. Two of the first set of OCW Scholar courses are mathematics courses: Single Variable Calculus and Multivariable Calculus. Applets were designed and built for use in the first of these courses.

Visitors to these sites are often working alone, so students must be able to operate and interpret these mathlets without instructor or peer support. This isolation results in some surprising requirements. For example, one student remarked on a chat site associated with the course:

The mathlets are clear, but when I use them for the exercises then I get confused. For example, when they say to find the slope of the tangent line, I don't know if they're talking about the red tangent line or the yellow tangent line.

The student was referring to the Secant Approximation applet; See Figure 1. This student does not yet distinguish between tangent lines and secant lines. The quote illustrates how easy it is for a student to experience a sense of understanding without actually comprehending what the visual is designed to convey. This gap in understanding is only revealed when the student is presented with an exercise that

Figure 1. The secant approximation applet

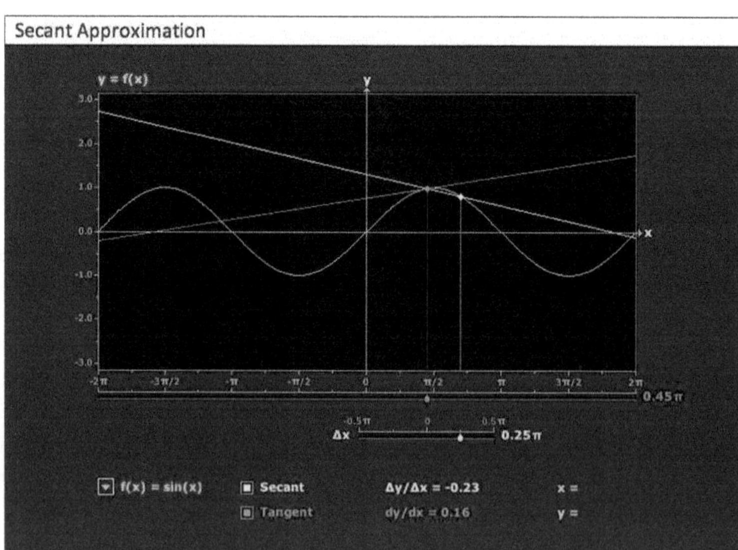

links the visual with a conceptual or computational task. Labels, color-coding, and layout do not by themselves induce an understanding of the intent of the applet. This is a universal characteristic of student response to applets, but it is perhaps especially severe for self-learners.

All the MIT Mathlets have a Help page, which details the various functionalities of the applet. As part of our response to the additional explanatory demands imposed by the OCW Scholar distance learning model, we have prefaced that text with a brief description of the conceptual content and intended use of the applet.

Design of the Secant Approximation Applet

The applet Secant Approximation (See Figure 1) illustrates many of the design features common to all the MIT Mathlets. This applet is designed to illustrate the convergence of secant lines to the tangent line at a point on a graph; this is a graphical image of the definition of the derivative of a function.

A menu offers a choice of five functions. We argue that a few well-chosen functions suffice to illustrate all the aspects of this particular curricular juncture which one is likely to want to explore. The advantage of limiting choice is that we have total control over the visual characteristics of the graph. This principle extends to all the applets in the MIT Mathlets collection.

The size and shape of the graphing window is independent of the choice of function, but the units vary. For trigonometric functions, the horizontal axis is marked off in multiples of π, while the vertical axis is in units. The aspect ratio is correct, however, so that a line of slope one makes a 45° angle with the x-axis. Failing to adhere to this kind of proportion can disrupt student comprehension.

The color cyan is used for three components in the applet: the choice of function, the graph of that function, and the label $y = f(x)$ of the graph. These are all representations of the same data, and so are displayed in the same color.

A slider below the graphing window allows the user to select a value of x. The slider position is indicated by a red dot, which is repeated on the graph above and connected to it by a thin vertical red line segment, to drive home the connection between these two marks. There is also a red readout of the x value to the right of the slider. Red indicates the position on the graph.

A yellow checkbox labeled [Secant] offers the option of displaying data connected with the secant line, all colored yellow: a slider, centered at the chosen value of x, allowing choice of Δx; a readout of the value of Δx; the secant line itself; a ghosted yellow line segment indicating the connection between Δx and

$x + \Delta x$; and a readout of the difference quotient $\frac{\Delta y}{\Delta x}$. The Δx slider is centered at the position of x. Sliders which slide is a novelty of this tool, and JAVA code executing it had to be added to the library.

One of the standing conventions, valid in all the MIT Mathlets, is that clicking on one of the numbered hashmarks sets the value of the corresponding parameter to that value. Important values are also artificially enforced: It may not be possible to station the Δx slider, for example, exactly at $\Delta x = 0$, because the computer screen is divided into discrete pixels. But careful programming guarantees that dragging the slider across $\Delta x = 0$ will result in the display appropriate to $\Delta x = 0$. This is one example of the many ways one needs to "cheat" in order to more accurately reflect the mathematical truth.

When $\Delta x = 0$, the readout for $\frac{\Delta y}{\Delta x}$ reads "Undefined". What secant line should be displayed when $\Delta x = 0$? The yellow dot slides behind the red one, and with no second point for the secant line, there is no secant line. It disappears. This is another standard convention of these applets: when a displayed object ceases to have meaning, it disappears.

Now is the time to take the limit. This is achieved by clicking on a second checkbox, labeled [Tangent]. Since it pertains to the given value of x, it is red. When it is checked, the tangent line at this point is drawn—in red, of course—and the value of the derivative is displayed, also in red.

A final functionality of this tool, again standard in the MIT Mathlets, is the creation of crosshairs over the graphing window centered at the cursor location (and suppressed by depressing the mouse key) and a readout of the values of x and of y under the cursor. The intent here is to allow the user to perform "experiments"—verifying the displayed value of the slope of the secant line, for example.

The Secant Approximation mathlet incorporates many visual cues to support student understanding of the relationship between secant and tangent lines. It is as simple and easy to use as possible.

Use of the Secant Approximation Mathlet

These applets can be used in a variety of ways. Experience shows that however charming they may seem to the professional, free exploration of them does not work well; the graphical medium is not intrinsically more conceptual than the symbolic one, and students tend to be timid in exploring and unsure what to look for. Individual students need a highly structured and detailed script. For groups one may be somewhat more open ended. We may use Secant Approximation to illustrate the potential usages of these applets.

Elementary Use

A straightforward way to get students involved with the MIT Mathlets is to invite them to make some measurements and then verify them computationally. We illustrate this approach using Secant Approximation. The main point of this applet is to help students grasp the trick needed to make sense of the tangent line at a point on a curve, here the graph of a function: You have to pick a different point on the graph, consider the "secant line" through those two points, and then look at the limit as the second point approaches the first. The elementary problems begin with that.

1. Select $f(x) = 0.5x^3 - x$ from the pull-down menu, and be sure that the [Secant] checkbox is checked and that the [Tangent] checkbox is unchecked. Set $x = 1.0$ (by clicking on the hash- mark, for example). Move the Δx slider and observe the secant line. Make a small table of values of $\frac{\Delta y}{\Delta x}$ as Δx ranges from –0.1 to +0.1. What is your prediction about the slope of the tangent line? Why is $\frac{\Delta y}{\Delta x}$ "undefined" (and why does the secant line vanish) when $\Delta x = 0$?
2. Now check [Tangent] and uncheck [Secant]. (There is no need for the Δx slider anymore, so it disappears!) Drag the x handle and observe that the red line is indeed always tangent to the cyan graph. Use the applet to estimate to within 0.01 the values of for x which $\frac{dy}{dx} = 0$. These are locations of extrema for the function $f(x)$!
3. Now compute the difference quotient $\frac{\Delta y}{\Delta x} = \frac{f\left(x + \Delta x\right) - f(x)}{\Delta x}$ by hand for this example, and take the limit as $\Delta x \to 0$. Does the calculation agree with the experimental prediction when $x = 1$? Does it agree with the observed values of *x* for which $\frac{dy}{dx} = 0$?

Intermediate Use

One of the attractive features of carefully constructed computer manipulatives, such as the MIT Mathlets, is that they often invite questions which elucidate the educational point at hand but which one might not have been able to ask without them. These can take the form of a game, and can greatly help the student in absorbing what the applet has to offer. Again, the Secant Approximation applet provides some examples.

1. Continue with $f(x) = 0.5x^3 - x$. Check both [Secant] and [Tangent]. Position the x slider at $x = 1.5$, and observe the range of values of the slope of the secant line as Δx moves across the range allowed by the applet. Record the extreme values. The difference between the extreme values gives an indicator of how small Δx has to be to obtain a good estimate for $\frac{dy}{dx}$. Now set $x = 0.5$ and $x = -1.0$, and make the same observations.
    a. What features of $f(x)$, near these three values of x, can explain the differences you observe?
    b. Sometimes the slope of the secant for $\Delta x > 0$ is greater than the derivative, and sometimes it's smaller. What features of $f(x)$ account for this?
2. With the same settings, take $x = 1.0$, where the slope of the tangent line is $f'(1) = \frac{1}{2}$. Make a table of values of $\frac{\Delta y}{\Delta x} - \frac{dy}{dx}$ for $\Delta x = -0.3, -0.2, ..., 0.3$. Use the quadratic approximation of $f(x)$ at $x = 1.0$ and this table to estimate the value of $f'(1)$, and compare this experiment with a hand calculation.

Advanced Use

The applets can also be used to explore issues beyond the immediate curriculum for which they were designed. In fact it has been our experience that these applets reveal phenomena we had not previously recognized, even in very elementary settings.

An example using Secant Approximation: Select $f(x) = \cos x$ from the pulldown menu, and check [Secant]. Select a nonzero value for Δx. What is the maximal value of the slope of the secant line, as you vary x? (This will be a function of Δx). (Answer: $\sqrt{2\left(1 - \cos\left(\Delta x\right)\right)} \Big/ \left|\Delta x\right|$).

Other Calculus Mathlets

We briefly mention several other innovative applets designed along the same principles, also for use in calculus courses.

- Creating the Derivative begins with the graph of a function (chosen from a pull-down menu). Checkboxes toggle displays of the tangent line, the graph of the derivative, the best approximating parabola, and the graph of the second order Taylor polynomial. The graph is drawn from some initial value of the independent variable up to some value which is controlled by a slider or by an animation; the graphs are "created."

- Graph Features uses sliders to control the coefficients of a cubic polynomial, which is graphed. Checkboxes toggle colored indications of regions where the graph is rising, falling, concave, or convex.
- Riemann Sums illustrates the geometric significance of a variety of versions of the Riemann sum, and illustrates the convergence properties of these approximations of the Riemann integral. One selects a function from a drop-down menu, and then selects a method using radio buttons from the options Evaluation point, min, max, trapezoidal, and Simpson's. In the case of Evaluation point, a slider opens by which one can choose where in each interval the function is to be evaluated. A second window illustrates the values of the corresponding sums as n increases. This tool can be very useful in helping students understand these choices, that for continuous functions they do not matter in the limit, and the improvement in rate of convergence offered by Simpson's rule.

Lessons Learned

Part of the challenge of assessing the effectiveness of online education is the difficulty of observing student responses directly. Consequently, we have little data on student reactions to the applets designed for self-learners. The majority of our information comes from students using the OpenStudy study group associated with our online Single Variable Calculus materials.

OpenStudy represents an attempt to replicate in distance learning the kind of help one gets from friends or study groups on campus. OpenCourseWare Scholar has provided links to OpenStudy and encouraged formation of OpenStudy groups studying the OCW Scholar courses.

The OpenStudy interaction differs from a campus interaction in many ways, however. On June 18, 2011, there were 418 students enrolled in the Single Variable Calculus study group, but there are rarely more than 3 students online simultaneously. So at any given time the level of activity on this site is less than would be found in a small group exercise in a classroom. On the other hand, questions persist and answers do too, and this somewhat compensates for a lack of immediacy. It also provides a valuable reference for students.

A review of questions posted to the study group reveals that students are more likely to discuss specific homework problems than general concepts A typical question is "What is the linear approximation of $\sin x$ at $x = 0$?"; questions like "How are tangent lines related to secant lines?" are rare.

Questions such as the student quote given in our introduction have led to improvements in the MIT Mathlets. For example, in an earlier version of the Secant Approximation applet, there was no toggle invoking the secant line; in fact the term

"Secant" never appeared on the applet. Adding the Secant checkbox (in yellow) increases the self-explanatory value of the tool, improves its logic, and encourages a better pedagogical sequencing.

No amount of on-screen information will make these tools self-explanatory for all or even most students, however. In their use on campus, it was found that using them in class helped students understand the significance of various elements and made it much easier for students to use them in homework assignments. An easy on-line analogue would be voice-over screen movies illustrating their use, and this is an enhancement which will accompany applets in subsequent OCW Scholar courses.

AEROSPACE ENGINEERING MATHLETS

In the spring semester of 2010, we developed an applet specifically for use in the Computational Methods in Aerospace Engineering course in the Department of Aeronautics and Astronautics at MIT. Although the mathlet was developed for this particular use, it is general to the pedagogy of eigenvalue stability (also known as absolute stability), an important concept for numerical methods courses in the advanced undergraduate curriculum (see M. Uri and Chen Greif, 2011;. Heath, 2002).

In our experience teaching undergraduate juniors and seniors in aerospace engineering, we have found that many students struggle with eigenvalue stability analysis, often focusing on the procedure instead of understanding the concept. With the Eigenvalue Stability applet, students are able to extricate themselves from laborious calculations and obtain real-time responses to variational modifications (e.g., by changing the time step or dominant eigenvalue magnitude for a given scheme). Whereas many of the MIT mathlets are self-contained and can be used to learn a mathematical concept from scratch, the eigenvalue stability mathlet is intended to accompany traditional instruction. Its primary role is to deepen understanding of the topic rather than to teach the process of the analysis. In the Computational Methods in Aerospace Engineering course, we used the mathlet as part of an in-class laboratory experiment after students were taught the mathematical procedures of eigenvalue stability analysis.

This section is organized as follows. First, we give some of the basic foundational material on eigenvalue stability analysis. It is necessary to understand the basic theory in order to appreciate the functionality in the applet. In the section that follows, we discuss the design considerations of the eigenvalue stability applet, emphasizing ease-of-use and maximizing the number of ways a user can interact with the applet. Then we discuss the use of the applet in three levels of difficulty (elementary, intermediate, and advanced) by drawing on lessons learned from an interactive laboratory assignment performed in the Computational Methods in Aerospace Engineering course in the spring semester of 2010. In the last section we present a few directions for future work.

Theoretical Background

Differential equations provide the language in which the laws of nature are expressed, and, consequently, the mathematics by which we understand the behavior of systems. In aerospace engineering, e.g., differential equations describe the unsteady fluid flow around an airfoil, permitting engineers to estimate lift and drag, quantities of interest that play an important role in aircraft design.

In most realistic applications, the differential equation does not admit solutions that are expressible in terms of elementary functions, and the only way to get an accurate understanding of the solution is via numerical methods. Often, the differential equation has many (thousands or more) degrees of freedom, e.g., the pressures and temperatures at many locations in the computational domain.

There is a well-established family of numerical integration schemes, and understanding their diverse applications is part of any engineering course in numerical methods.

Stability is one important consideration in selecting a numerical integration scheme[1] A numerical integration method is stable if the numerical approximation to the solution of the system of ODEs does not grow without bound. There are two forms of stability:

- **Zero Stability:** A numerical integration scheme is zero stable if the numerical solution is bounded as the number of time steps goes to infinity in the limit of zero time step.
- **Eigenvalue Stability:** A numerical integration scheme's application to a particular system of ODEs is eigenvalue stable if the numerical solution is bounded as the number of time steps goes to infinity for finite time step.

Eigenvalue stability depends both on the numerical integration scheme and on the governing equations to be integrated, in particular, through the associated eigenvalues of the linearization about an equilibrium. For a given numerical integration scheme, eigenvalue stability analysis leads to a stability criterion that restricts the choices of time step.

In this section, we sketch the analysis of eigenvalue stability for a linear system of equations with the forward Euler integration scheme and the midpoint method. This section is not intended to be a comprehensive primer on eigenvalue stability (see e.g. Uri & Chen Greif, 2011; Heath, 2002 for this), but rather to give the reader a sense of the mathematics involved and a greater appreciation of the discussion of the applet in the next section.

Consider the autonomous nth-order linear system of ODEs:

$$\dot{y}(t) = Ay(t), y\left(0\right) = y_0 \tag{1}$$

where $\dot{y}(t) = dy/dt$ is the vector of time-derivatives of the unknown variables $y(t) = \left[y_1, y_2, \ldots, y_n\right]$, and A is the $n \times n$ system matrix. Generic (non-defective) square matrices can be diagonalized. This implies that a change of variables replaces (1) with a decoupled set of scalar equations of the form

$$\dot{x}(t) \equiv f(t) = \lambda x(t), x(0) = x_0 \tag{2}$$

where $x(t)$ is the scalar unknown variable and λ is an eigenvalue (generally complex) of A.

In applying a numerical method, one chooses a time step Δt. Then one computes x_k numbers which approximate $x\left(k\Delta t\right)$ inductively, starting from the given value of $x_0 = x(0)$.

Example 1: Forward Euler Method

Let Δt be the time step and x_k be the numerical approximation to $x\left(k\Delta t\right)$ for each non-negative integer *k*. The forward Euler integration scheme approximates the time-derivative at x_k to first-order; i.e.

$$\frac{x_{k+1} - x_k}{\Delta t} = \lambda x_k \tag{3}$$

that is to say, $x_{k+1} = \left(1 + \lambda\Delta t\right)x_k$. Given x_0 and a time step Δt, one can calculate the approximate value of the unknown x at each time interval by iteration.

The multiplicative factor relating x_k to x_{k+1} is the *amplification factor.* It is a function $g(z)$ of the product $z = \lambda\Delta t$. In this scheme, $x_k = g(z)^k x_0$, which stays bounded provided $\left|1 + z\right| = \left|g(z) \leq 1\right|$. This imposes a restriction on time step Δt in terms of the eigenvalue λ. For a system of equations, this inequality must hold for all eigenvalues of the system matrix *A*.

This sets the pattern for more complex numerical schemes. The forward Euler is a single-stage numerical integration scheme. There are also multi-stage schemes that use data from more than one time step to estimate the solution at each step. The midpoint method is an example of a multi-stage scheme.

Example 2: Forward Euler Method

The midpoint method for $\dot{\mathrm{x}} = \lambda x$ uses the scheme

$$x_{k+1} = x_{k-1} + 2\Delta t \lambda x_k \qquad (4)$$

This is a homogenous second order difference equation. Its solutions will be linear combinations of solutions of the form $x_k = g^k x_0$ for suitable values of *g*. To find the, substitute this expression for x_k into (4) to see

$$g^{k+1} x_0 = g^{k-1} x_0 + 2\lambda\Delta t g^k x_0$$

Dividing by $g^{k-1} x_0$ we find a quadratic equation for *g*:

$$g^2 - 2zg - 1 = 0$$

The quadratic formula gives $g(z) = z \pm \sqrt{1 + z^2}$.

The amplification factor in this case is the multivalued function $g(z) = z \pm \sqrt{1 + z^2}$. In applying the midpoint scheme to (1), the time step Δt will satisfy eigenvalue stability provided that $z = \lambda\Delta t$ satisfies $\left|g(z)\right| \leq 1$ for both solutions to the quadratic equation and all eigenvalues λ of *A*. In this scheme, the region of stability is small: it is easy to see that $\left|g(z)\right| \leq 1$ if and only if *z* is purely imaginary of modulus at most 1. This scheme is therefore eigenvalue stable only for systems with purely imaginary eigenvalues. These features of the midpoint method are dramatically represented in the Eigenvalue Stability applet.

The study of the amplification factor for a variety of numerical integration schemes, governing equations, and time steps is a critical component in any numerical methods course in the undergraduate curriculum.

In the next section, we describe the design of the Eigenvalue Stability applet.

Design of the Eigenvalue Stability Mathlet

The primary design considerations for the Eigenvalue Stability Mathlet included the ability to understand the stability analysis from several distinct perspectives, options to toggle between a handful of well-known numerical integration schemes, and also the ability to control the position of one eigenvalue of the governing equation and the time step of the numerical integration.

There are three graphical perspectives from which eigenvalue stability can be viewed (Figure 2).

- In the upper right, the g-plane displays the unit disk, i.e., the stability region, and also the complex point z mapped through the amplification factor $g(z)$. The angle θ can be scrolled to outline the unit circle in the g-plane and observe the tracing of the corresponding stability boundary in the z-plane.
- The large graph on the left shows the location of the point $z = \lambda\Delta t$ in the complex plane and the stability region, which is the inverse map of the unit disk in the g-plane. The angle φ is the phase of the eigenvalue λ.

Figure 2. The eigenvalue stability applet

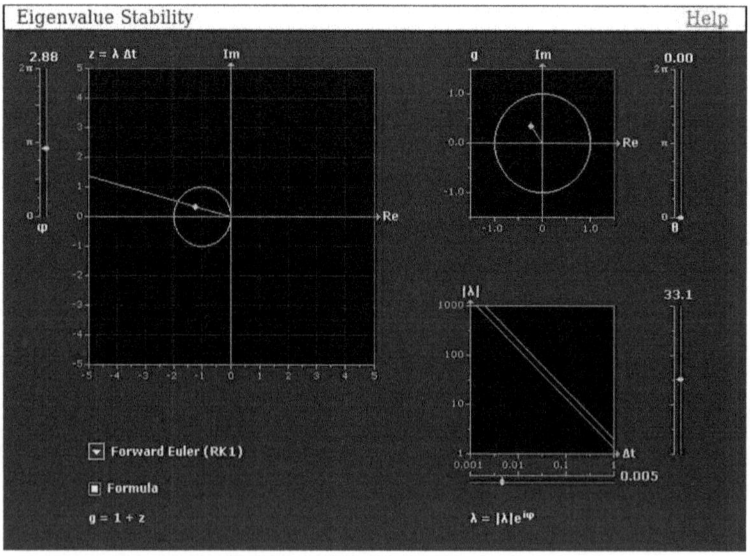

- Finally, in the lower right, a plot of $\left|\lambda\right|$ against Δt shows the shaded region corresponding to the points where $z = \left|\lambda\right| \Delta t e^{i\varphi}$ lies in the stability region of the z-plane.

All of the perspectives are dynamically integrated so that a modification in one plot is immediately reflected in the others.

The Eigenvalue Stability Mathlet illustrates these three perspectives and the relationships between them for six commonly used numerical integration schemes. For any of these methods, one can choose to show or hide the formula for the amplification factor with a radio button that appears just below the drop down menu for the numerical method.

One approach to using the applet is to begin by specifying an eigenvalue in the governing equations of interest. This can be performed as follows:

- Specify the magnitude of the eigenvalue using the scrollbar on the lower right plot. This produces an orange circle in the z plane indicating all of the values of z associated with that magnitude.
- Select the phase of the eigenvalue using the φ scrollbar to the left of the z plane. Once selected, an orange dot appears at the intersection of the circle and the ray defined by φ.
- Select a numerical integration scheme and find the range of time steps for which the integration will be eigenvalue stable by adjusting Δt until the orange dot moves outside the blue-shaded region.

Use of the Mathlet

In this section, we present a set of exercises of varying difficulty demonstrating how the eigenvalue stability applet can be utilized by students. The applet can be used as the primary apparatus in a series of laboratory experiments, can play an integral role in a homework assignment, or can be used in lecture to perform a real-time demonstration of the analysis of numerical methods. Some of the exercises discussed next were inspired by the laboratory worksheet used in class in the spring of 2010.

Elementary Use

A critical component of eigenvalue stability analysis, and one that has often been a source of confusion for students, is the mapping of a region or boundary in one complex plane to the corresponding region in another. Eigenvalue stability is determined by the magnitude of the amplification factor $g(z)$, but it is the z-plane that contains the scaled eigenvalues. Thus, one needs to determine the region in the z-plane corresponding to the unit circle in the g-plane. This complex mapping can be explored by students for the six numerical schemes available in the drop-down menu:

1. Select a numerical integration scheme from the drop-down menu and sketch the stability region in the z-plane by rewriting $g = e^{i\theta}$ and evaluating for several angles $\theta \in \left[0, 2\pi\right)$.
2. Grab the slider to the right of the g-plane plot and slide it from $\theta = 0$ to $\theta = 2\pi$. The dots on the g- and z-plane plots turn yellow and the stability boundary is traced out in both planes.
3. Verify the accuracy of your sketch of the stability region by comparing to the blue-shaded region in the z-plane.

Intermediate Use

In the numerical solution of ODEs, it is necessary to balance, or tradeoff, accuracy, stability, and computational cost. Many schemes will be stable and accurate for very small-time step, but then more time steps are required and therefore more computational resources will be used. For a set of governing equations and a given numerical integration method, it is therefore necessary to obtain the maximum time step for which the integration is stable.

1. Calculate by hand the eigenvalues of the system matrix.

$$A = \begin{bmatrix} 0 & 1 \\ -2 & -1 \end{bmatrix}$$

2. Select Forward Euler (RK1) from the drop-down menu.
3. For one eigenvalue, set the phase φ and magnitude $\left|\lambda\right|$ using the scrollbars in the applet.
4. Determine the maximum allowable time step Δt by dragging the Δt scrollbar in the lower right plot. Observe as z changes magnitude and eventually crosses over the stability boundary.
5. Does the maximum allowable time step change if you repeat the analysis with the other eigenvalue? Why or why not?

Advanced Use

Although eigenvalue stability is often binary (either the scheme is stable or not for a particular problem), the eigenvalue stability applet can be used to demonstrate some additional nuances in the application of numerical methods. Even when a numerical scheme is eigenvalue stable, it still may not be accurate for a given mathematical model. In this example we consider the simulation of a simple harmonic oscillator, e.g., an undamped pendulum.

1. Using the small angle approximation $\sin\theta \approx \theta$, the dynamics of an undamped pendulum (with well-chosen parameters) can be written $\dfrac{d^2\theta}{dt^2} = -\theta$. At time $t = 0$, the pendulum is at rest at $\theta(0) = \pi/6$. Rewrite this second-order ODE as a system of ODEs by introducing the two-dimensional state vector $y = \left[\theta, d\theta/dt\right]$.
2. Implement the trapezoidal scheme and the midpoint scheme to simulate the pendulum, e.g. in MATLAB, from time $t = 0$ to $t = 10\pi$. Determine by hand the eigenvalues of the system of ODEs you derived in (1), and use the applet to choose a time step to preserve eigenvalue stability for each numerical integration scheme. (Do not select a time step that is too small since computational resources are limited.)
3. What is the difference between the simulated responses? Which simulation is more consistent with the model of an undamped pendulum?
4. Analyze the eigenvalue stability for the trapezoid and midpoint schemes once again using the applet. In particular, consider the position of the orange dot in the g-plane. What is the effect of the amplification factor in this problem?

FUTURE WORK

Unfortunately, having used this applet only during one semester, we do not have much feedback from students. It is essential that we better understand students' abilities to learn eigenvalue stability through the use of our applet so that we can design additional teaching tools (lectures, homework problems, and laboratory assignments) and make improvements to the applet itself. In the future, we would like to integrate the eigenvalue stability applet into more of the numerical methods courses taught across other departments at MIT.

CONCLUSION

Carefully designed computer applets can be integrated into course work in a variety of contexts and subjects. We have detailed the design considerations and use of one applet used in a distance learning beginning calculus class and one used in a classroom workshop in computational methods in aerospace engineering. Other examples could easily be given. From OCW Scholar Single Variable Calculus we call attention particularly to Riemann Sums, and from engineering subjects Nyquist Plots and Bode and Nyquist Plots. The JAVA library developed to support these applets makes it quite easy to build new applets. As is often the case with educational software, the greater challenge is in using them effectively within a course.

REFERENCES

Ascher, U. M., & Greif, C. (2011). *A first course in numerical methods*. Philadelphia, PA: Society for Industrial and Applied Mathematics. doi:10.1137/1.9780898719987.

Haynes, M., & Upton, D. S. (2008). Computer manipulatives in an ordinary differential equations course: Development, implementation, and assessment. *Journal of Science Education and Technology*, *17*, 124–137. doi:10.1007/s10956-007-9058-2.

Heath, M. (2002). *Scientific computing: An introductory survey* (2nd ed.). New York: McGraw-Hill.

West, B., Strogatz, S., McDill, J. M., Cantwell, J., & Hohn, H. (2008). *Interactive differential equations.* Reading, MA: Pearson Addison Wesley. Retrieved from http://www.aw-bc.com/ide

ENDNOTES

[1] Accuracy is another important, but separate, issue. Indeed, accuracy of a numerical scheme does not imply that its numerical solution to an ODE will be eigenvalue stable.

Chapter 8
Applets for Mathematical Learning

Robert Terrell
Cornell University, USA

ABSTRACT

The purpose of this chapter is to describe dynamic interactive applets, which are available to all students, to assist in the understanding of sophomore-level engineering mathematics. The topics illustrated here with screen shots are the partial differential equations for heat conduction in one and two space dimensions, the wave equation and Laplace equation in two space dimensions, systems of three ordinary differential equations, and the display of vector fields in three dimensions. An applet is also shown for iterated function systems, although this is not usually part of the engineering sequence. Additional applets mentioned here and available online include the wave equation in one space dimension and two-by-two matrices. The purpose of these applets, all written by the author and freely available, is to help students experiment visually and dynamically with the mathematical concepts.

INTRODUCTION

It is common knowledge that abstraction is hard to understand if the learner is not equipped with the proper background. Many years of teaching engineering students have led me to conclude that mathematics students have this problem with partial differential equations and Fourier series, because in many cases they have little mechanical experience. For years when teaching the heat equation, for example, I described the heating and cooling of iron bars at the blacksmith's forge, and the changing of the colors with temperature, and this did not help the students very

DOI: 10.4018/978-1-4666-4050-4.ch008

much. I could talk about something closer to their experience, such as the cooling of pizza. This seemed to be more effective. With the wave equation it is possible to discuss the vibrations of a musical instrument in order to introduce Fourier series. This topic is very real to the string musicians in the class who have experienced the higher harmonics with their own eyes and ears. Even so, this discussion does not usually convey to all students the mental pictures which are needed for understanding the mathematics. For this reason, writing some dynamic software to help students experiment with the ideas of these equations in a visual manner seemed imminent. These are interactive java applets, and the main purpose is to let students explore the mathematics visually as a supplement to whatever textbook is being used. Very little background is required of the student. This chapter describes these applets. We show pictures from some of them, and discuss the mathematical context. Of course, one cannot see the full power of the dynamical environment unless the applets are actually utilized.

The Heat Equation in One Space Dimension

The heat equation describes the flow of energy due to conduction from hot regions to colder regions of static material. For material which can be considered one-dimensional, the temperature u is a function of time t and position x along an interval.

The typical undergraduate course shows how to solve the equation in one space dimension using separation of variables. Although this computation introduces many important ideas, it does not help all students to get a sufficient understanding of the meaning of the equation. So I have developed an applet which allows students to explore the possibilities.

In Figure 1, several screen shots are shown for a case in which the initial temperature is piecewise constant. The smoothing property of the heat equation is obvious when you watch it run. Specifically, the applet has a "Stop" button, but no matter how fast the student tries to click the button the solution already looks smooth. The boundary values in Figure 1 are 0 degrees at one end and -4 at the other. The temperature is displayed both as a graph and using colors to indicate temperature. Colors in the applet range from blue at the coldest parts to black to orange at the hottest. Because this is dynamic software, the student can see the colors and graph change with time. The comparison of the graph with the colors is also an important feature for students because most introductory mathematics texts emphasize the graphs of functions, while in many applications they ought to be thinking of the domain of the function and what is happening there. The colors are drawn in the material, the domain in which the heat conduction occurs. The student can explore both initial and boundary conditions in this applet. For example initial conditions can be typed

Figure 1. Screen shots from the heat equation 1D applet. The initial conditions shown at bottom are piecewise constant, and time increases upward in this figure. The applet is dynamic and the student sees the changes in real time and in color.

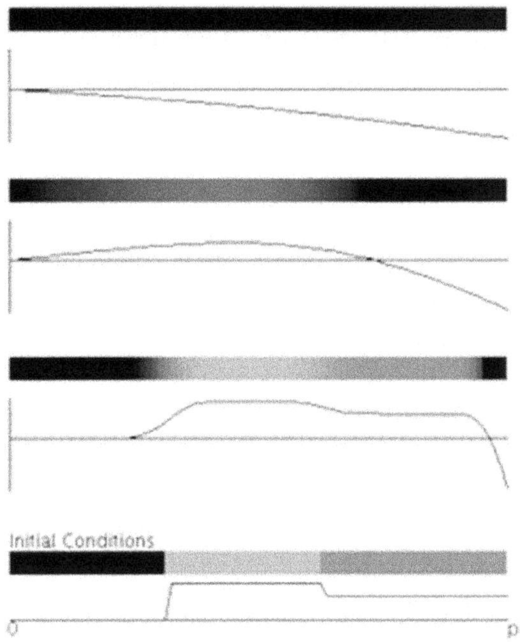

in as "u(x,0) = x*cos(2*x)" and boundary conditions may be used which are time dependent. Examples are "Left = 0", "Neumann Right" for insulation at the right end, "Right = 1/(1+t^2)".

To further explore the software, I present the following problem:

Problem 1: Suppose a metal bar has a hot spot near the right end initially, and the ends are maintained at 0 degrees. Then while the bar cools, does the hot spot move toward the end or toward the center of the bar?

The problem was suggested to me by Lars Wahlbin. It is a good example of a question related to ordinary objects, but it is hard to guess the correct answer unless you already know the mathematics, or have a lot of experience heating metal bars. The standard separation of variables approach resolves it by formulas, but the applet can resolve it visually before you know the formulas. For example, there is a solution u(x,t) = exp(-t)cos(x)+exp(-4t)cos(2x), which shows that higher spacial frequency is accompanied by faster decay in time. This implies that small features, such as a hot spot off center, will tend to die out, and visually the hot spot moves toward the center. The beginning student will not necessarily understand this argument right away, but the applet will demonstrate the concept quickly and dynamically. Perhaps the student will become motivated to understand the mathematical argument after seeing the result dynamically. See http://www.math.cornell.edu/~bterrell/h1.

The Heat Equation in Two Space Dimensions

Once the students have explored the one dimensional heat equation, they may be curious to see similarities and differences in the two dimensional case. There is then an applet for heat conduction in a rectangle. Again the initial and boundary values can be set by the user. All standard elementary functions are allowed, as well as the Heaviside function which was used in the initial conditions of Figure 1. The boundary conditions may again be time dependent, which is considerably more flexible than many texts discuss. There is an "Example" button on the applet, which cycles through a few built-in examples, but of course the purpose of the applets is to allow the student to try various things. All the examples can be edited and modified however the user wishes.

Here is a good problem for students to explore.

Problem 2: For the heat equation in two dimensions, suppose there are two blocks of steel with alternating 600 degree and 0 degree initial conditions as in Figure 2. For one of the blocks, the boundary temperature is maintained at 300 degrees. The second block is insulated all around the boundary. Then it is true that both blocks will approach an equilibrium temperature of 300 degrees. But which one will get there first, and will there be much difference in the rate of approach?

This problem statement is understandable in ordinary language, but it is hard to guess what happens. It can be resolved by trying the applet, or by trying a solution formula, or by thinking about the derivation of the heat equation. The derivation is not always done in elementary courses. If it is done, then the student knows that insulated boundary conditions, that is normal derivative zero at the boundary, mean that no heat energy flows across the boundary. But the 300 degree boundary can

Figure 2. A problem to explore using the heat equation 2D applet. Initial conditions consist of hot and cold areas, like a chess board. How does the evolution depend on the boundary conditions?

$$u_t = u_{xx} + u_{yy}$$

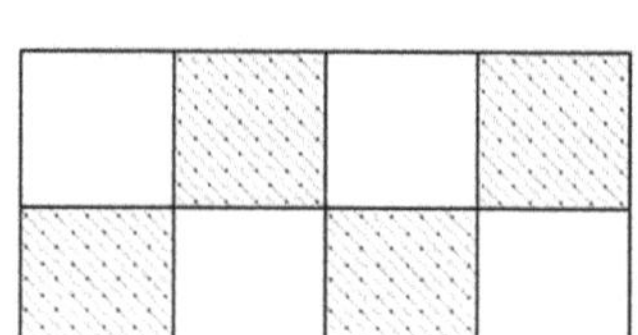

only be maintained by adding energy at the colder parts and removing it at the hotter parts. So it is almost certainly the case that the insulated block goes more slowly to the equilibrium.

This conjecture can then be tested in various ways. The argument leading to the conjecture may not be available to the student, if the derivation is not included in the course. Nevertheless the question can be explored using the applet; see http://www.math.cornell.edu/~bterrell/h2.

The Laplace Equation

The Laplace equation in two space dimensions may be viewed as the steady state case of the heat equation, so the heat 2D applet can also be used to display solutions of the Laplace equation. The user enters boundary conditions which do not depend on the time, and then allows the program to run for a short while until a steady state is established. There are many applications of the Laplace equation, so the student can interpret the display in the applet in ways unrelated to temperature. For example, there is the perfect fluid model which uses the gradient of u as the fluid velocity. Then the applet picture of a blue region near an orange region can be interpreted in at least two ways. If it is temperature, then the heat energy flows from orange to blue. If it is fluid velocity potential, then the fluid flows from blue to orange. In both cases, the flow is orthogonal to the level sets (See Figure 3).

Vector Fields in Three Dimensions and Ordinary Differential Equation Systems

Vector fields in three dimensions are difficult to visualize. We have two applets that can help. One is the vec applet (http://www.math.cornell.edu/~bterrell/vec) which displays vector fields, and the other is the de applet (http://www.math.cornell.edu/~bterrell/de) which solves systems of three ordinary differential equations. As with the PDE applets, the student can type the equations and initial conditions into the applets. In the case of de, the student can modify the stepsize and the length of time for which the integration is performed.

Figure 3. The heat equation 2D applet displays level sets of a harmonic function

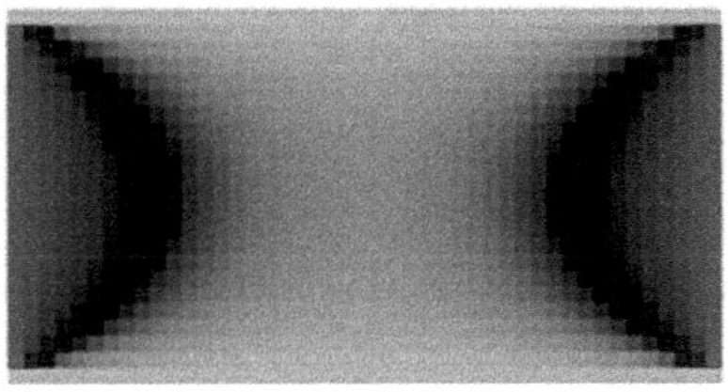

In the vector field applet the student is also able to add trajectories to the display, but unlike in the de applet the initial condition is chosen randomly. The emphasis is on the appearance of the field arrows. Figure 4 shows a vector field displayed from two different viewpoints. Three trajectories are shown. As with the other applets the student types whatever field is desired into text areas in the applet. The rotation is interactive, so the student controls the rotation and sees it occur in response to the mouse motions (See Figure 5).

The Wave Equation in Two Space Dimensions

In this applet (http://www.math.cornell.edu/~bterrell/w2) the student enters initial and boundary conditions for the wave equation in a square region. This is the most visually active applet, due to the nature of the wave equation itself. The student sees a lot of motion. Boundary values are allowed to be time dependent, constant, or Neumann type (See Figure 6).

Here is a problem which can help the student to understand product solutions of the form $u(x,y,t) = cos(at)sin(bx)sin(cy)$, which are found by separation of variables. It is not a hard question, but the student may know the solution formulas, and still be surprised at the visual comparison.

Figure 4. Three views from the de applet are shown. The student may type any system, and rotate the display interactively while the computation is in progress. This is the Lorentz system.

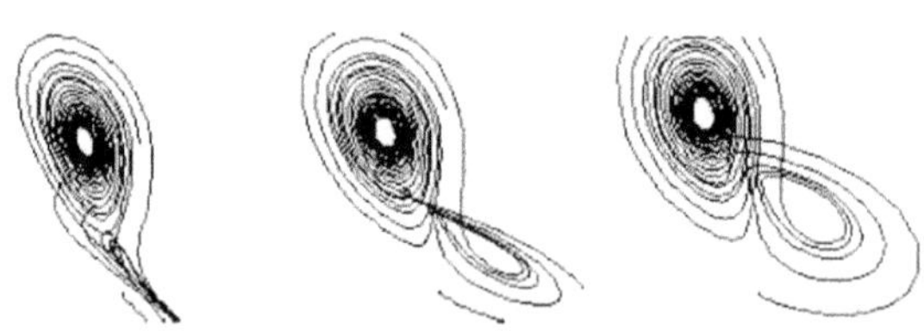

Figure 5. Two views of a vector field. Students can modify the field and rotate the viewpoint. The applet can also add trajectories of the field. Three have been added here.

Figure 6. The wave equation in two space dimensions. The example shown is more complicated than those in Problem 3.

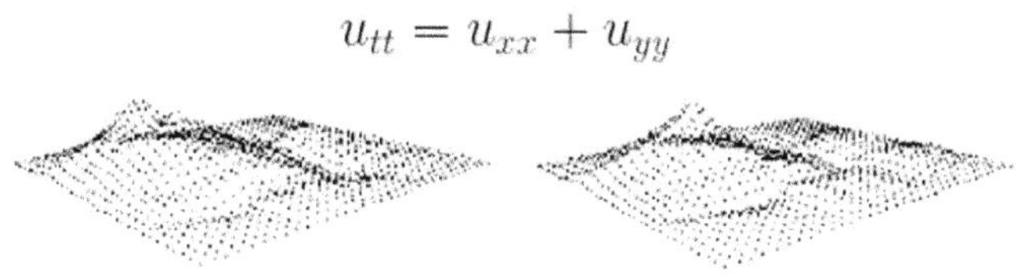

Problem 3: Two sets of initial conditions are *u(x,y,0) = sin(2x)sin(8y)*, and *u(x,y,0) = sin(4x)sin(4y)*, each with initial velocity 0. Which solution will oscillate faster?

The formula u(x,y,t) = cos(at)sin(bx)sin(cy) gives a solution when a^2 = b^2+c^2. In Problem 3, the values of a^2 would have to be 66 and 32, respectively, for the two sets of initial conditions. Therefore the first one oscillates about forty percent faster than the second one. This makes a big difference when it is seen dynamically.

Iterated Function Systems

This applet (http://www.math.cornell.edu/~bterrell/ifs) concerns material not usually included in engineering mathematics courses, but which can illustrate some matrix concepts which are included. It allows the user to enter several affine transformations of the plane. For each transformation, the user simply clicks three points which are to be the images of (0,1), (0,0), and (1,0). We can call each such set of three points a frame. The applet then draws a few thousand trajectories using Barnsley's (2000) algorithm, to display the limit of the system. What distinguishes this applet is that each frame point can be moved interactively while the iteration is taking place. This allows the user to see the changes dynamically. This produces such a variety of results that the student might forget to think about matrices and the contraction mapping theorem (See Figure 7).

Figure 7. Two iterated function systems. They differ only in the placement of one point in one frame.

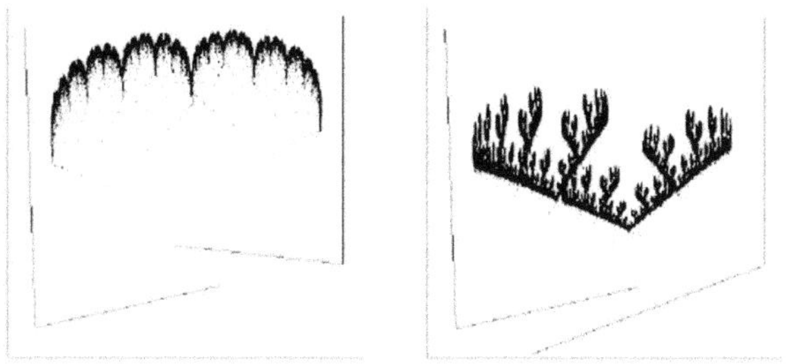

Other Applets

There is also a linear algebra applet (http://www.math.cornell.edu/~bterrell/la) for practice with two by two matrices, to learn about equation solving, transformations and eigenvectors, and a wave equation applet (http://www.math.cornell.edu/~bterrell/w1) for waves in one space dimension. These are standard material for engineering mathematics, and certainly ought to precede the ifs applet and the two dimensional partial differential equation applets. However, the applets require so little background that the student is free to explore them in almost any order. It is my intention that the student will be interested then to learn the mathematics during or after the exploration.

We have described software for learning mathematics, which is freely available to all students with access to a Web browser. The java source code is available online, for any educational use (Permission is not granted to use my software in any commercial product). It may be worth adding that one of these, the differential equations solver, is also available in a C source code version so that users with only a text interface may solve some differential equation systems. The standard disclaimer of any educational software applies here too: All of this software is for the education and enjoyment of students and their teachers, and should never be used for any other purpose.

CONCLUSION

The dynamic nature of some objects described by mathematics is brought to the students' attention by dynamic, interactive software, freely available.

REFERENCES

Barnsley, M. (2000). *Fractals everywhere* (2nd ed.). San Francisco, CA: Morgan Kaufman.

Terrell, R. (1995, Summer-Fall). Some software for teaching partial differential equations. *Newsletter for the Consortium for Ordinary Differential Equations Experiments,* 13-15. Retrieved from http://www.codee.org/library/newsletters/Summer-Fall%201995/

Chapter 9
Dynamical Software and the Derivative Concept

Ljubica Dikovic
Business Technical University, Serbia

ABSTRACT

Modern teaching trends impose the need of spending less time on the manipulative approach to differential and integral calculus, putting the accent on the conceptual understanding of the subject. This chapter presents the standard approach and method used to teach the derivative of a function and indicates some critical points in the teaching of the derivative, offering, at the same time, suggestions for overcoming them. As a supplement, the author gives e-resources that can make possible the implementation of a stimulating, visual, dynamic, and broadened method for teaching the derivative of a function.

INTRODUCTION

The National Council of Teachers of Mathematics (NCTM) Standards of America called for a mathematics curriculum that "emphasizes conceptual understanding, multiple representations and connections, mathematical modeling, and mathematical problem solving" (NCTM, n.d.).

The representation process includes the use of different models for organizing, memorizing and exchanging of math ideas with the aim of solving math problems and for a better interpretation of mathematics. Such models can be used for math-

DOI: 10.4018/978-1-4666-4050-4.ch009

ematics to be "seen, touched, presented" by use of multimedia materials, diagrams, graphic reviews or symbolic expressions. Representation should become an important supplementary element in teaching mathematics; so every teacher should choose a particular type and amount of representation materials to enrich the conventional form of a math lesson.

Representation—graphic and otherwise—in the function of reasoning has been explored by many researchers. One conclusion is that switching representations may often be a key to problem solving (Samdani, 2009). The standard representational forms of some mathematical concepts, such as the concept of function, are not adequate for students to construct the whole meaning and grasp the whole range of their applications (Monoyiou & Gagatsis, 2008).

The question, i.e. "How can a mathematical concept or relation be represented?", on the other hand, invites an answer to how a mathematical concept can be 'represented as' and it concerns our attempts to visualize the concept in a particular context for a particular problem solving task (Portides, 2008).

Finally, in order for future teachers to make an effective use of multiple representations in their teaching, they themselves need to experience and explore the potentials of technology as a learning resource rather than a computational device. As NCTM suggests, digital technologies provide visual models or representations that many students are unable to generate through their independent efforts (Ozmantar et al., 2010).

Calculus is the name for the symbolic calculation of the rate at which a function changes (differentiation). Derivatives and their applications are studied as a part of differential calculus which is one of the most important areas of mathematical analysis. In general, we can say that derivative is a measure of change, so its application helps in finding the area of monotony, concave monotony, finding the points of local extrema, and so on, which is extremely helpful in solving many engineering, financial and other real-life problems; Consequently, full mastery and understanding of the fundamentals of differential calculus is a requirement for every student.

BACKGROUND

The major obstacle to understanding the teaching of differential calculus is a large number of complex mathematical and dynamical concepts that the students did not encounter in their previous schooling. It is well known that students have great difficulty with the concept of the limit, derivative and integral, which are strongly linked. Currently many students are unprepared to study calculus, see no relevance in the topics taught, and fail the calculus course (Anderson & Loftsgaarden, 1988).

To alleviate these difficulties and obstacles, teachers often make the mistake of trying to reduce the teaching of differential calculus to a series of manipulative rules, which is still unacceptable for students, because it does not contribute to the fundamental, conceptual understanding of the matter in question. Why calculus cannot be made easy? What is the role of technology in teaching and learning calculus?

Modern teaching trends impose the need of spending less time on the manipulative approach to calculus, putting the accent on the conceptual understanding of the subject. The various technological tools (e.g., graphing calculators, Web-based mathematics applets, etc.) could be integrate in a high level mathematics course (e.g., Calculus) in order to stimulate visual, dynamic and broadened method for teaching the derivative of a function.

If mathematicians need to think visually, why do we keep such thinking processes from students? The dynamic nature of calculus should be in conjunction with a dynamic/interactive method of display. The Internet has many e-resources to help students get an intuitive feel and to visualize the derivative concept, allowing visual demonstrations and individual investigations into the mathematical ideas, which suggest new ways of approaching the derivative. This paper considers different ways in which student process information and focuses on the need to complement deductive thinking with different aspects of the derivative through numerical, symbolic, and graphical representations.

This form of learning is not a replacement for formal deduction, but a complement to it. It enables the less able student to grasp essential ideas that would previously be too difficult when framed in a purely formal theory.

The choice of the e-resources available on the Internet and used in this paper, simulates in an interactive, visual and animated way the very fundamentals of the differential calculus processes; the fundamentals which are, by their nature, dynamic and active. A flexible, graphic approach to calculus is based on our sensory perception of some physical activity of what we see changing and of how fast it changes, which is-inversely- related to the accepted symbolic meaning. That explains why the tools, chosen for presenting the process, are adequate for this teaching unit.

Analysis of the literature suggests that the impact of technology integration in the classroom is highly dependent on the manner in which it is used.

Some authors described a technology integration course planning assignment that was developed to enhance teachers' technological pedagogical content knowledge (Kersaint, 2007). For this reason, it is essential that teachers and researchers examine in particular how technology is used in teaching (Sheehan & Nillas, 2010).

The research literature provides supporting evidence that technology can enhance students' understanding of mathematics concepts (Rochowicz, 1996; Graham & Thomas, 2000; Garafalo, Drier, Harper, & Timmerman, 2000; Kersaint, 2003;

Kellogg & Kersaint, 2004; Ocak, 2005; Lim & Khine, 2006; Yaacob, Yuzita, & Wester, 2010) and improve their achievement. The new technological capabilities, including motion and connectivity, promise to transform teaching and learning in mathematics, making critical math concepts accessible to all students (Heller, Curtis, Jaffer, & Verboncoeur, 2005).

THE DERIVATIVE CONCEPT

Standard Teaching Approach

Classical applications that caused the development of differential calculus in the 17th century, dealt with finding velocity and acceleration, as well as the equation of the tangent line. Isaac Newton pioneered the development of differential calculus in the 18th century dealing with the problem of velocity:

Let $s = f(t)$ be the function describing the displacement of the point T by the line, where s is the distance, and t is the time. If we suppose that the displacement started from the point zero, that is, that $f(0) = 0$, then the point T by the time t_0 covers the distance $s_0 = f(t_0)$, and by the time $t > t_0$ the covered distance is $s = f(t)$ (Ljasko et al., 2007; Merkle, 2006).

The average velocity of point T in the given time interval (t_0, t) is $\dfrac{s - s_0}{t - t_0} = \dfrac{\Delta s}{\Delta t}$.

If f is a differentiable function, then when $t -> t_0$, the expression approaches the instantaneous velocity of point T at the time t_0,

$$v_0 = v(t_0) = \lim_{t->t_0} \frac{s - s_0}{t - t_0} = \lim_{t->t_0} \frac{\Delta s}{\Delta t} = f'(t_0).$$

Thus, velocity is the derivative (with respect to time) of an object's displacement (distance from the original position), and similarly, acceleration is the derivative (with respect to time) of an object's velocity.

Special accent in this part of unit presentation should be placed on the essential difference between two concepts: the average rate of change and instantaneous velocity (derivative).

If $f : D \to R$, $x \in D$, then the derivative of a function at any point x is defined in the following way (Newton's notation):

$$f'(x) = \lim_{\Delta x -> 0} \frac{f(x + \Delta x) - f(x)}{\Delta x}.$$

The important theorem is proved: If the function f is differentiable at point x_0, then it is also continuous at that point, i.e. $\lim_{x->x_0} f(x) - f(x_0) = 0, \quad \lim_{x->x_0} f(x) = f(x_0)$.

The function has no derivative at the points of discontinuity; so, the theorem is inversely invalid, which means that if the function f is continuous at the point x_0, it does not need to have a derivative at that point (for example $y = |x|, \quad y' = sign(x)$).

The example shows that the function is continuous, but the derivative of the function has at the point $x = 0$ a discontinuity of the first kind; so the function $y = |x|$ has at the point $x = 0$ the derivatives from the left and from the right. The example of this function is useful in the introduction of the concept of the left-sided and right-sided derivatives and also for connecting that concept with the points of the discontinuity of a function.

The question which requires a special explanation is: When is the function differentiable at some point, and when not?

Experience shows that this part of unit presentation is rather confusing, because the new concepts of the left and right domain of the point, the limiting value and increment, which are now introduced for the first time, have to be related to the previously studied concepts of continuity and discontinuity (Hohenwarter, Kreis, & Lavicza, 2008; Tall, 2009, 2010). That is why we suggest the use of computer animations, java applets and other graphic models, which can help make a visual image of processes otherwise difficult to be followed by symbolic notation.

Gottfried Wilhelm Leibniz, a philosopher and a mathematician, also developed, independently from Newton, the fundamentals of differential calculus in the 17th century, by solving the problem of finding the tangent of the curved line at some point.

Let a curve be defined by the function $y = f(x)$ (for example $f(x) = x^2$), where f is a differentiable function.

The secant of the function passing through the points $(x, f(x))$ and $(x + \Delta x, f(x + \Delta x))$, where $\Delta x \neq 0$, is the direction with the quotient $tg\alpha = \frac{f(x + \Delta x) - f(x)}{\Delta x}$.

Observing the change of the angle or the slope of the secant line (α) due to the decreasing of the increment of the argument Δx, we quickly come to the conclusion that the slope of the tangent line is the "limiting position" of the secant line, so the slope of the tangent line, which we call a derivative ($\alpha \to \alpha_0$, $tg\alpha \to tg\alpha_0 = f'(x)$),

is easily connected with the formula $f'(x) = \lim_{\Delta x \to 0} \dfrac{f(x+\Delta x) - f(x)}{\Delta x}$.

From this it is clear that the equation of the tangent line on the curve $y = f(x)$ at the point x_0 is $y - f(x_0) = f'(x_0)(x - x_0)$; similarly, the equation of the vertical line on the curve at the given point is derived from here.

For $y = f(x)$, the derivative can be thought of as the rate of change in y with respect to the rate of change in x. A derivative is a measure of how an output changes for infinitesimal changes in input. For a single-variable function, the first derivative is the slope of the function at that point, and it equals the slope of the tangent at that point. The derivative of a function at a chosen input value describes the best linear approximation of the function near that input value.

This part of the unit presentation should be followed by some animations showing the convergence of the secant to the tangent line, so the students can get a visual dimension of the dynamics of the process of change, and not only the knowledge of the concept itself.

The questions that need special attention are: Why is the tangent slope so important? What is a tangent line?

The basic teaching aim in this unit is to make the concept of the limiting value intuitive enough; that concept is motivated by the definition of the tangent line to the graph of $y = f(x)$ at the point $x = x_0$.

The special accent in this unit is on the fact that differential calculus studies the instantaneous of rate of change of continuous functions, which is called the derivative of that function and that, in a graphic sense, the slope of the tangent line at the given point represents the instantaneous rate of change, so it represents the derivative itself, while the secant represents the mean value of the rate of change; the fact that interconnects these two important graphic concepts with the concept of the derivative of a function.

It sometimes happens that some students manage to learn the set of routine, manipulative rules, and are able to solve even the complex derivatives, but they do not understand the very essence of the process of differentiation. By introducing computerized graphic animations, we try to overcome this serious problem in differential calculus teaching.

The idea of infinitesimal calculus develops with the introduction of the concept "x tends to a", offering the image of the quantity becoming arbitrarily small, but never equal to zero. The formal definition $\varepsilon - \delta$ of the limiting value which points to the very fundamentals of the process has many quantifiers, which is further confusing for the students; so because it is very difficult to calculate the limiting values by definition, the students readily accept the introduction of a manipulative rule.

A typical comment of mine student made about this was: "It is typical for the teachers to show us a lot of complicated ways before we come to the easy one to do it."

The question we should insist on is exactly the question of understanding the very fundamentals of the process of limiting value that lies at the bottom of the process of function derivative.

The choice of e-resources available on the Internet, which we will present in the following part of the paper, simulates in an interactive, visual and animated way the very essence of the fundamental processes of differential calculus; the essence which is dynamic and active by nature. That explains why our choice of the tools needed for the presentation of the process is the right one for this unit.

The Use of E-Resources for Intensifying Conceptual Understanding of Function Derivative

E-resources are suitable because they enable multiple representations of mathematical concepts. By making the tables while moving along the curve or by zooming in critical points, students will be able to process information in a more interesting way; also by working with different aspects of the same concept they will develop critical thinking and find the answers to the questions of the type " what-if'. A flexible graphic approach to differential calculus is based on our sensory perception of a physical action of what we see changing and the speed of that change; which is inversely connected with the adopted symbolic meaning.

Therefore, the concept of a tangent line and a secant line has to be completely mastered, as well as the concept of the growth of the independent variable and the function; that must be done before we start practicing the limiting values of the functions. Graphs describe the relations among the changing quantities. The concept of variables as changing quantities should be properly mastered. The difference between dependent and independent variables is crucial. Function, as a logical relation between variables, should also be mastered well.

In the introductory part, the numerical approach to a graph is studied, with the aim of connecting the concept of the continuous rate of change with the slope of the tangent of a curve at a certain point. Students can really calculate the rates of changes by drawing tangents on curves; the most important in that is to learn how to make the difference between the instantaneous and the mean rate of growth change.

The function derivative for some known functions can be checked using computer packages or calculators.

Graphical analysis of derivative of functions could be illustrated on the following examples:

- The applet for secant and tangent lines (SLU, n.d.) is designed for the visual exploration these lines. In particular it allows the user to compare the graph of a function with the right secant at values x_0 and $x_0 + dx$, or with the left secant at values $x_0 - dx$ and x_0.
- The applet for slopes and derivatives is a GeoGebra applet that traces out the slope of the line tangent to a curve as the base point is moved along the curve. For example, the calculations of the slope of a tangent line $f(x) = ax^2 + bx + c, a \neq 0$ (with the parameters a, b, c controlled by sliders) for different values and the drawing of these slopes to the independent variable, will give the line $y = 2ax + b$, as a gradient function. The main viewing window gives the graph of a function f(x), at point T, the tangent line at point T, and a point with that turns the slope into the height of a point (See Figure 1) (Dikovic, 2009).

The possibility of making flexible changes during a graphic interpretation, as on the example of the tangent slope taken as a basis for calculus, allows students to get the knowledge of the rules, and even helps make the proofs less rigid. The GeoGebra applet illustrates the geometric interpretation of the derivative. The student can enter a formula (or parameters) and the applet will graph an arbitrary function. Moving the sliders lets the student explore families of functions. The actual graph of the derivative is revealed point by point as the user

Figure 1. The applet for visualization the slopes and derivatives

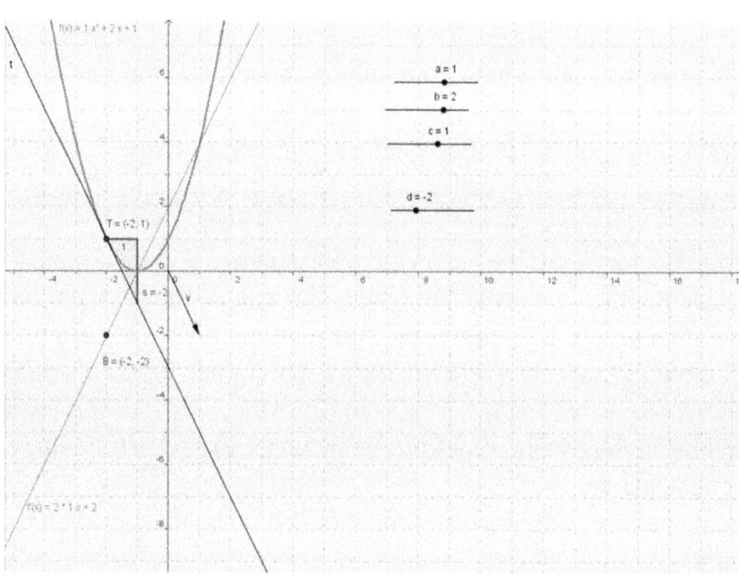

drags a slider. At the same time, the linear element representing the tangent line and its slope is dynamically shown.
This applet is designed to build the intuition between the idea of a derivative being the slope of the line tangent to a curve and the idea of a derivative as a function symbolically computed form the formula of the original function.

- **Applying the derivative:** Let the students analyze some practical and interesting problem (IES, n.d.): "A man is riding on the surf. We set $f(x)$ as the curve of the wave. Observe the slope of the surfboard. The trace of the slope is the derivative of $f(x)$." By solving the given problem, students actually apply derivatives for tackling a real life problem. In this case, the Java applet was used to simulate a real-life situation.
- **Understanding the graph of the first and second derivative:** The following task can be defined: Observe and analyze a parallel graph of any function, $y' = f'(x)$ and $y'' = f''(x)$ and notice the following conclusions (See Figure 2) (Sarra, n.d.):
  At any point x where the function f(x) is decreasing (increasing), the slope of the tangent line is negative (positive), so $y' = f'(x)$ is negative (positive). In the extrema, the tangent line is horizontal; the slope is 0, so $y' = f'(x) = 0$.
- **Recognizing the graph of the original function, and the graph of the first and second derivative:** After working through these materials, the student should be able to recognize graphically which of three graphs is the graph of the original function, the graph of derivative of the function and the graph of the second derivative of the function.

Figure 2. The applet for understanding the graph of the $f'(x)$ and $f''(x)$

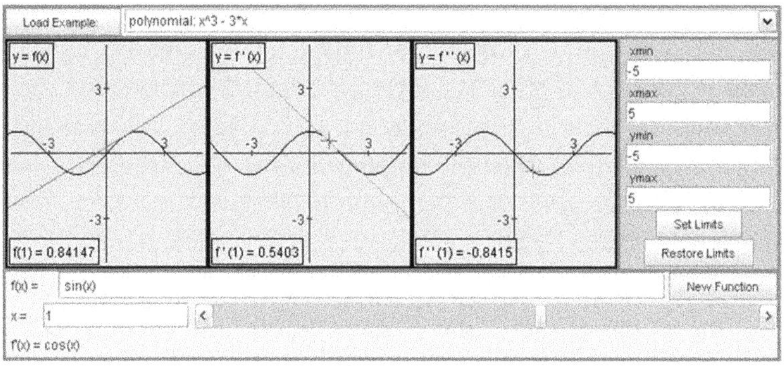

Figure 3. The applet for recognizing the graph of the $f(x)$, $f'(x)$ *and* $f''(x)$

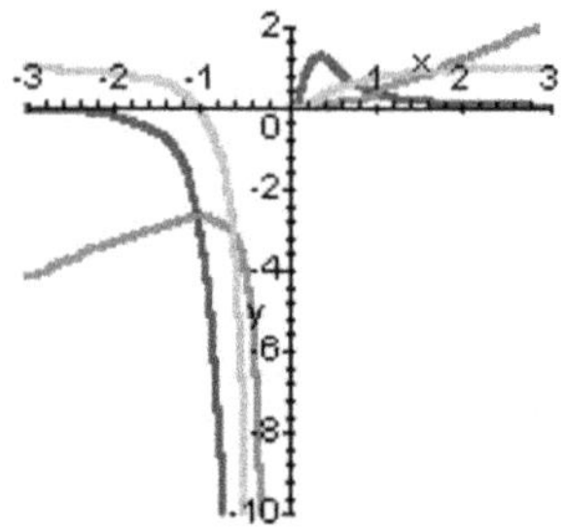

In the beginning, the instructor should let the students prove graphically that the gradient of the function is sine cosine, or that the following is the graph of the function $f(x)$, its derivative $f'(x)$ and second derivative $f''(x)$ (the slider moves synchronized points on the three graphs) (See Figure 3) (UTK, n.d.):

Student should be able to answer the following important questions: "Which graph is the graph of $f(x)$, $f'(x)$, and $f''(x)$?"

Students can be able to check their answers: "As the blue function appears to head towards zero (or to almost cross the x-axis), the green function starts to flatten out (or to almost change slope). When the green function has a negative slope, the blue function has negative values, and when the green function has a positive slope, the blue function has positive (above the x-axis) values. So the blue function is the derivative of the green function. When the green function crosses the x-axis, the red function changes from a positive slope to a negative slope. And when the red function has a negative slope, the green function has negative (below the x-axis) values, and vice versa. So the green function is the derivative of the red function. Hence, the graph of $f(x)$ is red, $f'(x)$ is green and $f''(x)$ is blue".

Testing for Convexity and Concavity of a Single Variable Function

A function that is convex is often synonymously called concave upwards, and a function that is concave is often synonymously called concave downward. For a twice-differentiable function *f*, if the second derivative, *f* ''(*x*), is positive (or, if the acceleration is positive), then the graph is convex (or concave upward); if the second derivative is negative, then the graph is concave (or concave downward). The points - where concavity changes - are called inflection points.

Students can explore the following assertion: "The graph of a continuous function is concave up, if the second derivative is positive." in a visual way. Sometimes this rule is easily remembered and correctly applied, but without real understanding. In this case, if the second derivative is understood as a rate of change of an angle of a tangent line, the slope is increased; that is why the curve must be concave up. Using visual descriptions based on the observation of how the change in one variable influences the other variable, the rules can be accepted as generalizations. Taking the derivative to be a dependent variable for every value of the independent variable of a continuous function, will lead to a new, gradient function which describes how the original function changes in every given value of an independent variable.

The use of multiple representations to calculate the derivative in a graphic, algebraic and numerical way provides the conditions for better understanding of the concept. However, it is absolutely necessary for students to communicate clearly between the method and the conclusion. Students may be asked during the exercises to complete the sentences in order to explain their methods or ways of work or to interpret their results.

Or, if they need to find the local minimum of the function, students are expected to know how to use differential calculus and to be able to show certain mathematical steps which could lead to the solution. The graphic calculator is a powerful research tool, but we should be aware of the fact that research is not always a mathematical solution. For example, research including the graphic calculator can lead to an analytical solution; and after the solution has been found, the graphic calculator can be used for checking it.

In summary, students should understand the definition of a derivate as a limit as two points of a function get infinitesimally close; the relationship between differentiability and continuity; graphical, numerical and analytical representation of derivatives, and how they are interpreted as an instantaneous rate of change. Technology provides us with tools that can make understanding of these challenging concepts more accessible to learners.

CONCLUSION

Knowing how to use different dynamical software does not mean to be able to understand the derivative concept. It means that graphic software could be used for visualization of a concept and for the understanding of the fundamentals of the process, and especially for visualization of the connection between a graph and an algebraic representation of a concept.

As the unsatisfactory level of knowledge of the results in the final exam on the Calculus course, there is a need to invent new methodologies, using modern features and quality of applied ICT. All teachers should be open to learning new technologies, implementing them strategically and effectively in mathematical teaching contests. In this way, technology could play an important educational role, changing the focus from mechanical and repetitive processes to the comprehension of meaning of derivative concepts as well as instruments that enable the modeling, discovering and understanding of many different kinds of situations.

REFERENCES

Albers, D. J., Anderson, R. D., & Loftsgaarden, D. O. (Eds.). (1986). Undergraduate programs in the mathematics and computer sciences: The 1985-1986 survey. Mathematical Association of America, 7.

Diković. (2009). Applications GeoGebra into teaching some topics of mathematics at the college level. *The International Journal of Computer Science and Information Systems*, *6*(2).

Diković. Ivković, & Milivojević, M. (2010). *Visualizing and understanding limits via internet technologies*. Paper presented at the 3rd International Conference Science and Higher Education in Function of Sustainable Development. Užice, Srbija.

Garafalo, J., Drier, H. S., Harper, S., & Timmerman, M. A. (2000). Promoting appropriate uses of technology in mathematics teacher preparation. *Contemporary Issues in Technology & Teacher Education*, *1*(1). Retrieved from http://www.citejournal.org/vol1/iss1/currentissues/mathematics/article1.htm.

Graham, A., & Thomas, M. (2000). Building a versatile understanding of algebraic variables with a graphic calculator. *Educational Studies in Mathematics*, *41*(3), 265–282. doi:10.1023/A:1004094013054.

Heller, J., Curtis, Jaffer, & Verboncoeur. (2005). *Impact of handheld graphing calculator use on student achievement in algebra 1*. Heller Research Associates.

Hohenwarter, M., Hohenwarter, J., Kreis, Y., & Lavicza, Z. (2008). Teaching and learning calculus with free dynamic mathematics software GeoGebra. In *Proceedings of the 11th International Congress on Mathematics Education*. University of Nuevo Leon.

IES. (n.d.). Retrieved from http://www.ies.co.jp/math/java/calc/doukan/doukan.html

Kellogg, M., & Kersaint, G. (2004). Creating a vision for the Standards using online videos in an elementary mathematics methods course. *Contemporary Issues in Technology and Teacher, Education*, *4*(1). Retrieved October 14, 2007, from http://www.citejournal.org/vol4/iss1/mathematics/article1.cfm

Kersaint, G. (2003). Technology beliefs and practices of mathematics education faculty. *Journal of Technology and Teacher Education, 11*(4), 549–577. Retrieved from http://www.editlib.org/p/2222.

Kersaint, G. (2007). Toward technology integration in mathematics education: A technology-integration course planning assignment. *Contemporary Issues in Technology & Teacher Education, 7*(4). Retrieved from http://www.citejournal.org/vol7/iss4/mathematics/article1.cfm.

Lim, C. P., & Khine, M. S. (2006). Managing teachers' barriers to ICT integration in Singapore schools. *Journal of Technology and Teacher Education, 14*(1), 97–125.

Ljasko, Boljarcuk, Gaj, & Golovac. (2007). *Zbirka zadataka iz matematicke analize.* Belgade, Serbia: Nasa Knjiga.

Merkle. (2006). *Matematicka analiza*. Belgarde, Serbia: CET.

Monoyiou & Gagatsis. (2008). The stability of students' approaches in function problem solving: A coordinated and an algebraic approach. In *Proceedings of the Conference of Five Cities: Nicosia, Rhodes, Bologna, Palermo, Locarno "Research in Mathematics Education"*. University of Cyprus.

NCTM. (n.d.). Retrieved from http://www.nctm.org/

Ocak, M. (2005). Mathematics teachers' attitudes toward the computers. *The Turkish Online Journal of Educational Technology, 4*(3).

Özmantar, M., Akkoç, H., Bingölbali, E., Demir, S., & Ergene, B. (2010). Pre-service mathematics teachers' use of multiple representations in technology-rich environments. *Eurasia Journal of Mathematics, 6*(1), 19–36.

Portides, D. (2008). Representation per se vs. representation as: A useful distinction for mathematics education. In *Proceedings of the Conference of Five Cities: Nicosia, Rhodes, Bologna, Palermo, Locarno "Research in Mathematics Education"*. University of Cyprus.

Rochowicz, J. A. Jr. (1996). The impact of using computers and calculators on calculus instruction: Various perceptions. *Journal of Computers in Mathematics and Science Teaching, 15*(4), 423–435. Retrieved from http://www.editlib.org/p/15206.

Sarra, S. (n.d.). *Derivatives*. Retrieved from http://www.scottsarra.org/applets/calculus/Derivatives2.html

Sheehan & Nillas. (2010). Technology integration in secondary mathematics classrooms: Effect on students' understanding. *Journal of Technology Integration in the Classroom*, *2*(3).

SLU. (n.d.). Retrieved from http://www.slu.edu/classes/maymk/SecantTangent/SecantTangent.html

Tall. (2009). Dynamic mathematics and the blending of knowledge structures in the calculus. *ZDM – The International Journal on Mathematics Education, 41*(4), 481–492.

Tall. (2010). *A sensible approach to the calculus*. Paper presented at the National and International Meeting on the Teaching of Calculus. Puebla, Mexico.

UIUC. (n.d.). Retrieved from www.cs.uiuc.edu/homes/rsamdan2/Rajhans.MSThesis.pdf

UMN. (n.d.). Retrieved from http://www.ima.umn.edu/~arnold/graphics.html

UTK. (n.d.). Retrieved from http://archives.math.utk.edu/visual.calculus/3/graphing.3/index.html

Yaacob, M., Yuzita, S., & Wester, S. (2010). Towards the development of an automated learning assistant for vector calculus: Integration over planar regions. *The International Journal for Technology in Mathematics Education*, *17*(2), 81–86.

Chapter 10
Supporting the Development of College-Level Students' Conceptions of Statistical Inference

Maria Meletiou-Mavrotheris
European University Cyprus, Cyprus

ABSTRACT

The transition from descriptive to inferential statistics is a known area of difficulty for students taking introductory statistics courses. This chapter shares the experiences from a teaching experiment in a college-level introductory statistics classroom that implemented an informal, data-driven approach to statistical inference using the dynamic statistics software Fathom© as an investigation tool. Findings from the study indicate that the informal inferences on which instruction focused in the first part of the course helped students develop understandings of fundamental aspects of inferential and argumentative reasoning that served as foundations for the formal study of inferential statistics in the latter part of the course. The affordances offered by the tool for delving deeply into the data to make sense of the situation at hand were instrumental in supporting student understanding of both informal and formal inferential statistics.

DOI: 10.4018/978-1-4666-4050-4.ch010

INTRODUCTION

Statistics, the science of learning from data, is divided into two main areas: descriptive statistics and inferential statistics. Descriptive statistics is devoted to organizing, summarizing, and presenting data. It involves using tabular, graphical, and numerical techniques to analyze and describe a dataset. Inferential statistics, on the other hand, is intended to reach conclusions that extend beyond the immediate data, to deduce that observed patterns in the data at hand are also present in some broader context. It improves decision-making in a variety of real-world situations by providing tools that enable the drawing of causal inferences, or inferences to populations using sample-based evidence.

Although statistical inference is the cornerstone of modern statistical concepts and methods, grasping the key ideas related to inferential statistics is a known area of difficulties for students (Green, 1982; Rubin et al., 1990; Garfield & Ahlgren, 1998; Gordon & Gordon, 1992; Rubin et al., 2006). Research in statistics education has long suggested that students have difficulty using inferential statistics methods appropriately in applied problems. For example, research on introductory college-level statistics courses suggests that even students who can successfully implement procedures for hypothesis testing and parameter estimation are often unable to use these procedures appropriately in applications (e.g. Gardner & Hudson, 1999; Reichardt & Gollob, 1997). As Erickson (2006) points out, "inference is so hard that even professional researchers use it inappropriately" (p. 1).

Traditionally, introductory statistics courses adopt a linear, hierarchical approach to the different statistical topics encountered in the course. The structure of almost every introductory statistics course is to first start with descriptive and exploratory data analysis, then move into probability, and finally go to inferential statistics. Statistical inference is presented in the statistics classroom as a set of formal tests and procedures, through which information contained in sample data is used either to estimate the values of the respective population parameters (i.e., construct confidence intervals), or to check claims made regarding the values of population parameters (i.e., perform hypothesis testing). The research literature indicates major student difficulties in the transition from descriptive to inferential statistics (Green, 1982; Garfield & Ahlgren, 1998; Gordon & Gordon, 1992). After having investigated the teaching and learning of statistical reasoning for several years, Rubin, Bruce, and Teney (1990), came to the conclusion that grasping the basic concepts of sampling and statistical inference is extremely hard for students. Understanding, according to the researchers, seems to break down as soon as non-determinism enters the classroom.

Recent advances of technology provide schoolteachers and college instructors with new tools for adopting informal, data-driven approaches to statistical inference that can help lay the conceptual groundwork for formal inferential reasoning (Rubin et al., 2006). The appearance, in particular, of dynamic statistics learning environments (e.g., TinkerPlots© (Konold & Miller, 2005) and Fathom© (Finzer, 2001)), which are designed explicitly to facilitate the visualization of statistical concepts, provides an enormous potential for making inferential reasoning accessible to students. These new technological tools provide a medium for the design of activities that integrate experiential and formal pieces of knowledge, allowing students to make direct connections between physical experience and its formal representations (Meletiou-Mavrotheris, 2003; Paparistodemou, Noss, & Pratt, 2008). Students can experiment with statistical ideas, articulate their informal theories, use the theories to make conjectures, and then use the experimental results to test and modify these conjectures. This leads to a shift in the focus of statistics from learning statistical tools and procedures (e.g., graphical representations, numerical measures) towards more holistic, process-oriented approaches that go beyond data analysis techniques. Statistics can be presented as an investigative process that involves formulating questions, collecting data, analyzing data, and drawing data-based conclusions and inferences (Franklin et al., 2007).

This chapter shares the experiences from a teaching experiment that implemented an informal, data-driven approach to statistical inference using the dynamic statistics software Fathom© as an investigation tool. The study sought to answer the following questions:

1. How does college-level introductory statistics students' reasoning about statistical inference develop from simple forms (informal) to more complex ones (formal), in a learning environment structured around the "big ideas" of statistics, and focused on providing ample opportunities for data-based inferences?
2. How can the affordances provided by a dynamic learning environment for formulating and justifying data-based inferences, be utilized in instruction to foster powerful notions of both informal and formal statistical inference?

BACKGROUND

Developing students' informal ideas of inference is a topic of current interest to many statistics educators, who have acknowledged the fact that despite their difficulties with the formal methods of statistical inference, students do have some sound intuitions about data (Rubin et al., 2006; Bakker, 2004), which can be refined and moved towards reasoning that has inferential qualities (Rubin et al., 2006).

According to Watson (2008), informal statistical inference represents a continuum of experience from the point when students start to pose questions about datasets to the point when they are about to meet formal inferential statistics. Along the way from informal to formal inference, a number of important ideas are added to the student package. Rubin et al. (2006) define informal inference as reasoning that involves consideration of the following related ideas:

1. Properties of aggregates rather than properties of individual cases.
2. Sample size and its effect on the accuracy of population estimates or on process signals.
3. Controlling for bias.
4. Tendency, distinguishing between claims that are always true and those that are often or sometimes true.

Informal inferential reasoning has been highlighted in a number of studies as a potential pathway for deepening learners' understanding of statistical processes and outcomes. Ben-Zvi (2006) links informal inferential reasoning to argumentation and the need for data-based evidence. Deriving logical conclusions from data is accompanied by the need to provide persuasive arguments based on data analysis. Integration and cultivation of informal inference and informal argumentation seem to be essential in constructing students' statistical knowledge and reasoning in rich learning contexts (Ben-Zvi, 2006).

Zieffler et al. (2008) view informal reasoning about statistical inference as the way in which students build connections between observed sample data and unknown or theoretical populations, and the way they make arguments or use evidence to support these connections. Building such connections between sample and population lies at the heart of informal statistical inference (Pratt et al., 2008). Makar and Rubin (2009) regard informal inferential reasoning of statistics as a reasoned but informal process of creating or testing generalizations from data that extend beyond the data collected. They consider the following three principles to be essential to informal inference:

1. Making generalizations (predictions, parameter estimates, conclusions) that extend "beyond the data".
2. Using data as evidence for these generalizations.
3. Using probabilistic language in describing the generalizations, including references to levels of certainty about the drawn conclusions.

Although the emergence of studies specifically focusing on informal inferential reasoning has begun to shed some light on this important aspect of statistical reasoning, research on the topic is still at an early stage. Since the investigation of informal inferential reasoning is a very recent endeavor, research "is only just grappling with understanding the conceptual building blocks for informal inferential reasoning as a pathway towards formal statistical inference", both in terms of the discipline and in terms of student cognition (Pfannkuch, 2006).

The current study contributes to the emergent research literature on informal inferential reasoning by describing findings from an instructional experiment designed to explore the ways in which the affordances of the dynamic statistics software Fathom© for formulating and justifying data-based inferences could be utilized to enhance the development of key concepts related to statistical inference. By investigating the interactions with Fathom© of a group of college-level introductory statistics students that made use of the software to clarify the process of drawing inferences from samples, the study explores the different ways in which student understandings of both informal and formal inferential statistics may develop as a result of interacting with this technological tool.

Features of the Dynamic Statistics Software Fathom©

Use of technology is essential to learning statistics. Technology can illuminate key statistical concepts by allowing students to focus on the process of statistical inquiry – on the search and discovery of trends, patterns, and deviations from patterns in the data, and on the communication of findings to others. Choosing the right software however is of paramount importance. Ben-Zvi (2000) argues that statistics instruction ought to employ the use of technological tools which support active knowledge construction, provide opportunities for students to reflect upon observed phenomena, and contribute to the development of students' metacognitive capabilities. Computer-based learning environments should be such as to encourage students to: (1) practice data analysis with an exploratory, interactive, open-ended working style, and combine exploratory and inferential methods, graphical and numerical methods; (2) make extensive use of multiple linked representations and simulations to construct meanings for statistical concepts and ideas; and (3) construct models for statistical experiments, and use computer simulation to study them (Ben-Zvi, 2000). Often, software employed in statistics classrooms do not posses these qualities. They are either professional statistical tools or simplified versions of professional tools that have been developed top-down (Biehler, 1995), i.e. from the perspective of the statistician rather than bottom-up from the perspective of statistical novices. Professional statistics software employed in the introductory classroom are often complex and not suitable for all students (Ben-Zvi, 2000), while the simplified versions of professional tools are simpler only in that they have fewer options (Konold, 2002).

The educational software Fathom© differs from conventional statistical packages in that it is a pedagogical tool designed for concept development (Finzer, 2000), rather than software designed for professional data analysis. Although a complete data analysis tool, Fathom©, which targets secondary and higher education students, was specifically designed to enhance learning and make statistical thinking accessible to students. Its design drew on current constructivist theories of learning and several years of academic research about the way students learn and process statistical concepts and the main difficulties they face.

Fathom© belongs to the new family of educational software in the teaching of statistics that came to be known as dynamic software, which offer an environment that permits the construction and flexible usage of multiple data representations. All of the software's objects are continuously connected and, thus, selection of data in one representation means the same data are selected in all representations. Changes in a data point in one representation are reflected in all related representations. Thus, students can interact with the data and see the immediate impact that their actions will have on the different representations of the data on the screen.

Fathom© combines dynamic capabilities with core features such as the "drag and drop" interface and the Formula Editor. The "drag and drop" interface allows students to drag rather than choose things from menus or dialog boxes. Dynamic dragging can have profound effects on student learning since "there is a qualitative difference between comparing two disparate states of a system and actively controlling the transition between those two states" (Finzer & Erickson, 1998). The Formula Editor, which has true mathematical formatting, allows students to write commands for calculating values, plotting functions, and controlling simulations, and to see the impact of these commands on the different data representations. Abstract mathematical ideas can be presented on the screen in symbolic and iconic forms. Several research studies have shown that linking abstract ideas with concrete representations facilitates learning (e.g. Wilensky, 1993; Pratt, 1998).

Dynamic statistics software such as Fathom© (and more recently Tinkerplots©, a software of a similar nature to Fathom© that was designed specifically for younger learners) do much more than producing fancy graphs; they facilitate the discovery of patterns through exploratory data analysis. Strong research support exists for the efficacy of dynamic computer graphics as instructional media that support active construction of knowledge by learners rather than forcing them to accept information provided by the computer without deep processing (e.g., Yu & Behrens, 1994; Hoyles & Noss, 1994; Meletiou-Mavrotheris, 2003; Bakker, 2003; Paparistodemou & Noss, 2004; Stylianou & Meletiou-Mavrotheris, 2005; Meletiou-Mavrotheris, Paparistodemou, & Stylianou, 2009). Attributes like the ability to link multiple representations, to provide immediate feedback, and to transform a whole representation into a manipulable object, have affordances towards more constructive pedagogical approaches than traditional packages. The direct manipulation of mathematical objects and synchronous update of all dependent objects facilitates learning by allowing users to ask "what if…?" questions, make conjectures, and then easily test and see these conjectures in action (Ben-Zvi, 2000). Technology goes far beyond the role of mere means for data display and visualization to become a tool for thinking and problem solving.

Several researchers have, in recent years, been exploiting the affordances provided by dynamic learning environments for promoting learners' ability to reason and argue about data-based inferences, with very encouraging results (e.g. Bakker, 2004; Ben-Zvi, 2006; Gil & Ben-Zvi, 2011; Meletiou-Mavrotheris, 2003; Rubin et al., 2006). Some of the conducted studies have demonstrated that even young children can develop powerful notions about inference when using appropriate data visualization tools (e.g. Pratt, 2000; Stohl & Tarr, 2002; Ben-Zvi, 2006; Paparistodemou & Meletiou-Mavrotheris, 2008; Paparistodemou & Meletiou-Mavrotheris, 2010). Paparistodemou and Meletiou-Mavrotheris (2008), for example, report on how a group of third grade students (8-year-old) formulated and evaluated data-based inferences using the dynamic statistics data-visualization environment TinkerPlots©. Ben-Zvi (2006), describes how the features of the dynamic software Tinkerplots©, in combination with a carefully designed learning trajectory based on growing samples heuristics, supported a group of 11-year-old students in deriving persuasive arguments based on data analysis. Stohl and Tarr (2002) describe how two average-ability sixth-grade students (11-year-old) were able, through using a variety of microworld tools, to formulate and evaluate inferences based on simulation data.

In the current study, Fathom© was employed in the hope that use of the software would promote conceptual change in college-level, introductory statistics students, and would support student understanding of both informal and formal inferential statistics.

CASE DESCRIPTION

Methodology

Study Design

A case study design was employed in the study. It was judged that this research strategy was well suitable to exploring, discovering, and gaining insight into the perceptions, actions and interactions with the dynamic software Fathom© of introductory statistics students. The study was exploratory in nature, and thus its purpose was not to prove or disprove hypotheses, but rather to generate descriptions based upon in-depth investigation of students' interactions with the technological tool, and of the impact this might have on their inferential reasoning. These descriptions, while of limited generalizability, may be used to understand similar situations (Stake, 1995) and can inform future research.

Context and Participants

The case studied was a college-level introductory statistics classroom in Cyprus. The author was the course instructor. The class met twice weekly, for an hour and-a-half each time, over a period of four months. There were 37 students in the class, all majoring in a business related field of study. Only a few of the students had studied mathematics at the pre-calculus level or higher.

The study examined the ways in which students' understanding of statistical inference developed from the simplest forms (informal) to the more complex ones (formal), in a technology-based learning environment carefully designed to promote inferential reasoning. The class meetings took place in the computer lab. Students worked individually or in small groups on a set of computer-based activities aimed at eventually building their understanding of inferential statistics using the software as an investigation tool.

Instruments, Data Collection, and Analysis Procedures

The study attempted to obtain a detailed description of how learners' initial informal understandings about statistical inference evolved into formal statistical reasoning, and to document the different ways in which the instructional tasks and technological tools employed promoted the development of inferential reasoning in students. In examining students' learning progress, a variety of both qualitative and quantitative data gathering techniques were used: classroom observation, videotaping of all computer sessions, videotaped interviews of selected students, and student work samples.

During students' engagement with the computer-based activities, the instructor assumed the role of a participant observant. She was interacting with the students, and was probing in order to get a better idea of their thinking processes. The nature of her involvement changed as the course progressed. While at the beginning her interaction with students was often instruction-based and she would, when deemed necessary, step in to help students reach solutions to their problems, at the latter part of the course she let students work on the activities independently. She would only step in either to offer technical assistance when necessary, or to probe more deeply into the thinking behind students' actions and discussions.

The videotapes collected during the course were first globally viewed and brief notes were made to index them. The goal of this preliminary analysis was to identify representative parts indicative of students´ approaches and strategies when performing specific tasks using Fathom© and of the ways in which the software influenced student thinking. The selected occasions were transcribed and viewed several times. We carefully studied and analyzed students' talk and actions. The method of analysis involved inductively deriving the descriptions and explanations of how the students interacted with the software and approached selected ideas of statistics. We attempted to verify or refute the conjectures inherent in the design of Fathom© by investigating the degree to which the features and structure of the software influenced students' approaches to statistical investigations, and how this, in turn, affected their individual schemes for key statistical concepts related to statistical inference. These interpretations were corroborated by the insights gained from examining the data collected from other sources. The descriptions derived formed the study findings that are outlined in the next section.

Nature of Instructional Experiment

Instruction aimed at motivating learners and helping them make connections between formal inferential reasoning and their prior experiences, beliefs, and understandings. It fostered a supportive and engaging learning environment, in which students were actively involved in constructing their own knowledge, through their own experiences and participation. Students actively engaged in the process of learning and developed their statistical thinking through authentic educational activities (projects, experiments, etc.), computer explorations, group work, and discussions.

Rather than presenting statistical content as a sequenced list of curricular topics, instruction focused on building conceptual understanding of the big picture of statistics. It emphasized and revisited a set of central statistical ideas, making them explicit throughout the course. Through open-ended projects and other activities, students learned where these ideas apply and how. In order to build bridges between procedural and conceptual perspectives, curriculum was designed based on a "progressive formalization" approach (NRC, 2000). Students first encountered the big ideas related to statistical inference informally, and then revisited them again and again as instruction moved to more formal understandings. Through data explorations of real or simulated data, instruction in the first part of the course focused on building a teaching pathway towards formal inference by helping students experience and develop the "big ideas" of informal inference: (1) knowing why they should compare centers, (2) describing and interpreting variability within and between sample distributions, (3) developing their sampling reasoning, and (4) knowing how to draw acceptable conclusions based on informal inference (Pfannkuch et al., 2004). Later in the course, students were introduced to confirmatory or formal inference methods, and began comparing empirical probabilities with the theoretical ones. They learned the formal procedures for building sampling distributions, constructing confidence intervals, and conducting hypothesis testing.

Assessment was aligned with learning goals. It focused on understanding of key ideas and not just on skills, procedures, and computed answers, and was an integral component of instruction (Franklin et al., 2007). The learning environment provided students with multiple opportunities for feedback and revision. Most of the hands-on and technology-supported instructional activities expected students to first make conjectures about the expected results, then to test these conjectures through computer simulations or other data explorations, and finally to reflect on and to evaluate on their results.

A core element of the instructional experiment was the dynamical statistics environment Fathom©. Instruction employed the features of Fathom© to help students visualize abstract concepts, explore data, to formulate and evaluate conjectures based on data, and to check conditions underlying inference procedures (Franklin et al., 2007). The activities were carefully designed to support but, at the same time, also explore students' evolving meanings.

The next section presents three teaching episodes that are indicative of the types of activities in which students engaged during the course, and of the ways in which the features of dynamic tool might have contributed to the construction of more powerful meanings regarding key ideas related to statistical inference.

SATs and GPAs: Classroom Activity

In this whole class-activity (taken from Erickson, 2001), which took place early in the course, students looked at gender differences in Scholastic Assessment Test (SAT) scores and Grade-Point Averages (GPAs) for 1000 first-year college students and tried to figure out how meaningful those differences were, using Fathom© an exploration tool. The SAT Reasoning Test is a standardized test for college admissions in the United States, and the GPA is an indicator of a college-level student's academic performance (it is a calculation of the average of all of a student's grades).

The instructor first asked students to look at the Fathom© summary table shown in Table 1, and describe what they noticed based on the table alone. Some students in the class argued that men do better on the math part of the SAT, while females have a higher FYGPA (first-year GPA). One student claimed: "Males are better in math," and another one added: "Females have a higher GPA because they study more, but males are smarter."

Unlike the students who were ready to give causal explanations for the differences in the means, several other students argued that knowing mean scores is not adequate information for making comparisons between males with females. We

Table 1. Summary table of math SAT scores and FYGPA

SATGPA		math	FYGPA
sex	F	521.942	2.54459
		484	484
	M	564.593	2.39607
		516	516
Column Summary		543.95	2.46795
		1000	1000

S1 = mean ( )
S2 = count ( )

need to look at the actual dataset, one student noted, because we need to find "things like the standard deviation." The instructor agreed with the need for looking more closely at the data and taking spread into account. She pointed out that the explanations some students gave for the reasons behind the difference in means make claims, which go beyond the information provided simply by comparing mean scores. She added that what they would now do to check their conjectures was to retrieve the actual data and look more closely at it, analyzing it using Fathom©.

After opening the file containing the dataset, students first drew a histogram of math SAT scores (Figure 1-left). The instructor asked:

Inst.: *Do you consider the distribution to be normal or not?*
Savvas: *Close to normal.*
Inst.: *In statistics we use the words close, fair enough a lot. In real world situations, one cannot expect to get perfect normal distributions. This is actually a good example of a real dataset following an approximately normal distribution.*

Next, students drew a histogram of the math scores separated by gender (Figure 1-right), and compared males and females using the new information provided by the graphs. Most noted that the two graphs are very similar and "*although males are slightly higher than females they are close.*" Then, students compared FYGPAs by drawing graphs such as the one shown in Figure 2. They noted that this time it is females who outperform males, with more females than males having a FYGPA of 3 or higher, but that again there is a lot of overlap. In order to get more insight, students compared results of males and females on the verbal part of SAT and saw that they were similar.

The instructor then asked students to go back and change their graphs to boxplots. Comparing the boxplots of the three sets of scores (See Figure 3), students saw that sex-related differences are more noticeable on math SATs and on FYGPAs than on verbal SATs.

Figure 1. Histogram of math SAT scores for all students combined (left), and separated by gender (right)

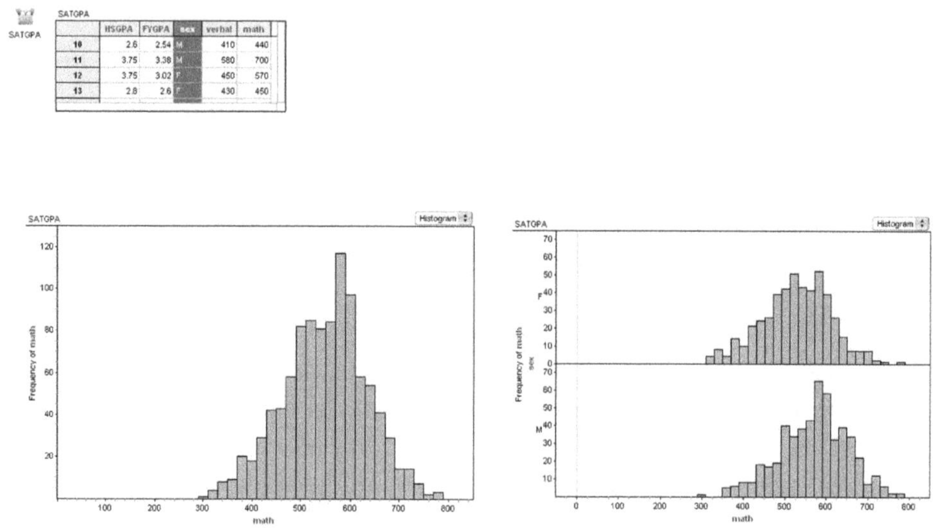

Figure 2. Histogram of FYGPAs separated by gender

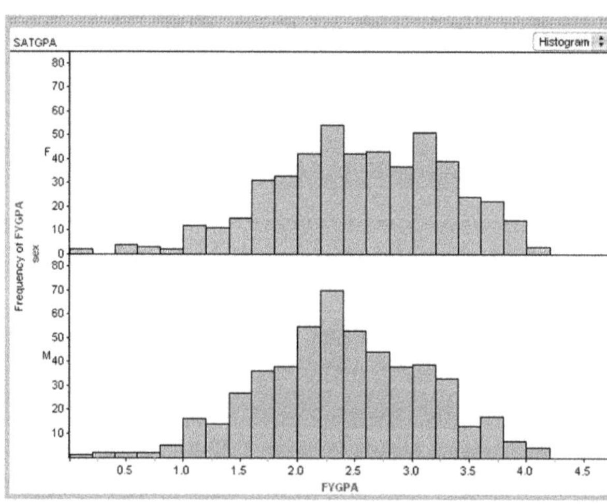

Next, the instructor asked students to explain how they knew, by looking at the graphs, that the difference in math scores is more meaningful (not just bigger) than the difference in verbal scores. One student noted that, looking at the boxplot of math SATs, one sees that the difference in scores between males and females is "a sizable proportion of the interquartile range." Another one thought the difference in math is more meaningful "because the box is not exactly lined up for the math

Figure 3. Boxplots of FYGAP scores (left), verbal SAT scores (center), and math scores (right) separated by gender

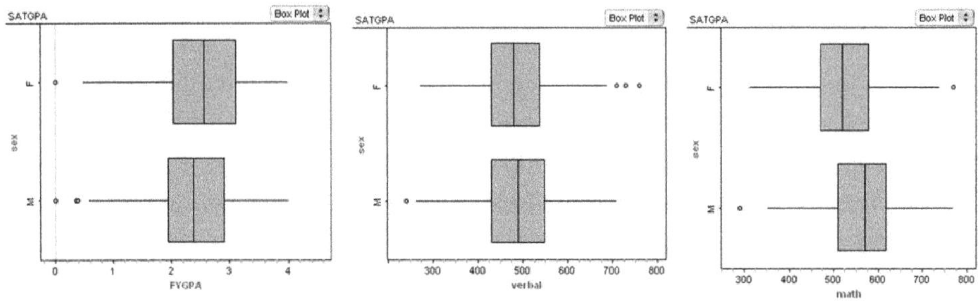

scores like it is for the verbal scores." One student remarked: "Q2 (second quartile) for males is almost equal to Q3 (third quartile) of females, showing that the differences in math scores are much greater."

One of the students made the assertion that by comparing the SAT and FYGPA distributions, we see that "Males are better than females in overall since they are much better in math." Another student challenged this student's assertion by stating that "female inferiority in mathematics is a social stereotype". The instructor agreed with this student's comment, commenting:

In the past women lagged behind men in mathematics achievement, however, this is no longer the case. International studies of mathematics achievement conducted in recent years such as the TIMSS study show that in most countries, including Cyprus, there is essentially no difference in performance between boys and girls.

It is true, the instructor pointed out to the class, that based on the displays, it does seem that in this dataset of US student scores males did in general have a higher math SAT scores than females, but at the same time there is a substantial overlapping of distributions. She added that even if the student's claim seemed plausible regarding the specific group of students from whom the data were collected, examining a graph is not adequate when conducting a statistical analysis; one also needs numerical summaries to support inferences drawn by looking at displays. She explained to the students, that later in this course they would learn statistical tests that would allow them to decide whether the observed differences among two distributions are actually significant or not. Moreover, she pointed out, it would be important to compare these data, which concern students in the US several years ago, with more recent data, since there might have been changes in student performance in SAT and FYGPA in terms of gender.

The purpose of engaging students in the "SATs and GPAs" activity was to help them understand two "basic lessons of statistics" which are the foundation of sophisticated inferential statistical methods: (1) one cannot compare measures of central tendency without taking variation into account; and (2) we can assess the salience of a between-groups difference by comparing it to the within-group variations (Erickson, 2001). Indeed, the activity served its purpose as the analysis of the responses students gave in the worksheets, and of the videotaped class session indicated. Although at first several students drew inferences and generalizations based only on the differences in means, through the activity they came to see that knowing mean or median scores is not adequate information for making comparisons between groups and that "we need to look at the actual datasets to find things like the standard deviation."

The "SATs and GPAs" activity is a representative example of the type of activities employed in the early part of the course in order to develop students' informal ideas of inference. Instruction provided an exploratory data analysis learning environment, which offered many opportunities for students to analyze real data and to draw conclusions based mainly on comparison of data distributions displayed graphically. This helped them to move away from "uni-dimensional" thinking and to become able to integrate center and variability into their analyses and interpretations of data. While at the beginning of instruction students tended, when analyzing or interpreting data distributions, to concentrate on their center and to ignore or under-recognized the variation around that center, during the teaching experiment there was a gradual improvement in their intuitions about variation and its effects. The culture of the classroom, in the activities it structured and the interpretations it valued, sent clear messages about the importance of variation in statistical reasoning and helped students overcome their inclination to interpret the world in deterministic terms. The change in student perceptions was a slow, non-linear process. It was not easy for students to shift from a deterministic view of reality to one that incorporates variation and uncertainty. Nonetheless, the emphasis of the course on the omnipresence of variation did help them make some steps beyond their deterministic thinking and start developing their informal inferential reasoning.

Similarly to what Ben-Zvi (2003), it was found that students' development of reasoning about variation was accompanied by a somewhat parallel development of global perception of a data distribution as an entity with typical characteristics such as shape, center, and spread, and sample size. Through their continuous participation in a variety of interesting computer and hands-on activities that elicited conceptions of variability and difference rather than centers and sameness, students "discovered the richness and complexity of data" (Hammerman and Rubin, 2003). They began to appreciate the fact that there is lots of variability in the real world, but also that we are often still able to detect patterns in the data (Shaughnessy, 1997). Eventually students began to see patterns in the variability of different phenomena, and thus shifted from focusing at individual values of a distribution to adopting a more distribution-oriented view of data (Bakker, 2003).

In dealing with distributions of samples, they began to get some idea of what is likely and what is unlikely to occur by considering the entire distribution of outcomes and to develop some sense of the effect of sample size on spread.

Coin Toss Activity

Before being introduced to the formal methods of statistical inference, students engaged in several activities inquiring them to create informal confidence intervals from distributions and decide, by looking at graphical displays, what range of values of that parameter they would consider reasonable, and which values seemed implausible and begged for another explanation (Erickson, 2001). One of first activities students engaged in was the "Coin Toss" activity.

This activity is an inference problem based on a simple binomial situation. Students were asked to decide how likely it is to get 40 heads when tossing a coin 50 times. They worked on this activity in pairs. Before moving to Fathom©, they made predictions as to what outcomes are likely when tossing a coin 50 times. Then, they used the software to model the experiment to find out how likely it was for 40 heads to have occurred by chance. A description of how two of the students, George and Anna, approached the task is next provided.

George and Anna first made the conjecture that getting 40 heads out of 50 throws of a coin simply by chance "could happen but it is unusual". Then they moved to the computer to test this conjecture. They first collected a single sample of 50 coin tosses by building a Table with 50 *cases*. They created the attribute *CoinToss,* and entered the command RandomPick("Head", "Tail") in the Formula Editor to simulate the tossing of the 50 coins:

George: *Let's do the Coin Toss problem now. What do you think we should do?*
Anna: *We will make a new table, this one we'll call CoinToss, now do...put the formula...RandomPick("Head", "Tail"*
Instr.: *Great. What do you want to see here?*
George: *New cases...should I put 50?*
Anna: *Yeah.*

The two students got the table shown in Figure 4-left. When prompted as to why they created this table, Anna explained: "Out of 50 tosses…You want to see how many heads you will get…." She went ahead to "drag" a graph on which "dropped" the *CoinToss* attribute, in order to get a bar graph of the outcomes (See

Figure 4-right). Pushing Control-Y re-randomizes the values in the table. George and Anna pushed Control-Y several times and saw how the bars of the graph, and the numbers in the table, would change. This helped them get an idea as to what values were likely. They wrote down the number of correct recognitions that came up each time. They noted that "40 never appears".

After students had repeated this re-randomization several times, the instructor let them know that Fathom© has a feature, the *Measures Collection*, which actually allows the user to record statistics – in this case the count of heads. They added the attribute *CountHeads*, which they specified by using the command Count(CoinToss= "Head") and checked that it gave them the same number as the number of heads displayed on the bar graph. They then used the *Collect Measures* command, which automatically collected the counts of five repetitions of the experiment (Figure 5).

Figure 4. Case table and bar chart for the "coin toss" task: sample collection of 50 coin tosses

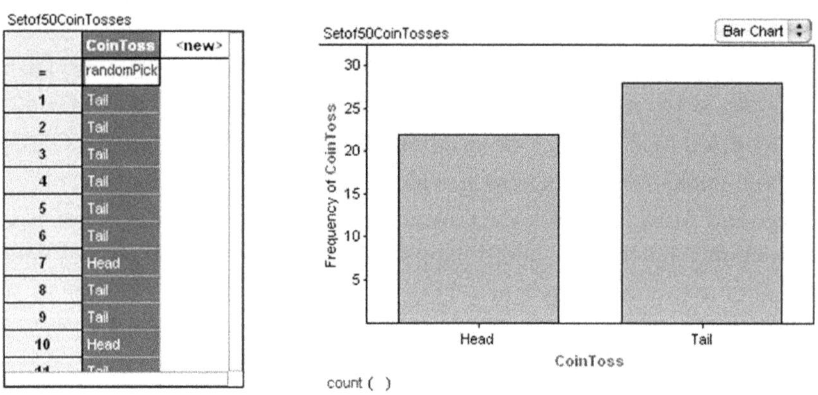

Figure 5. Case table and histogram of sample statistics for the "coin toss" task: a measures collection of 5 samples

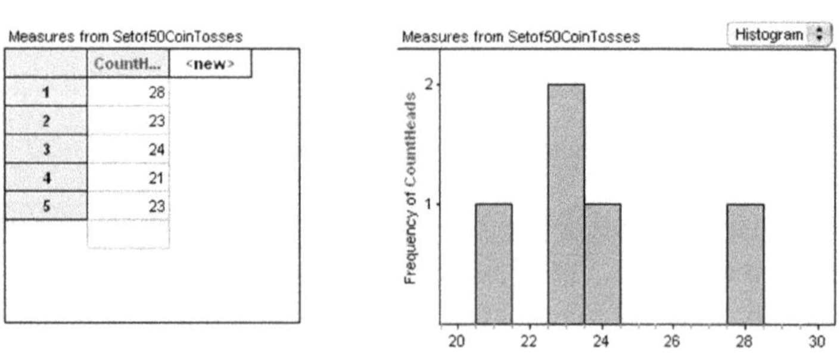

Before continuing, the instructor checked to ensure that they understood what these counts stood for:

Instr.: *What do you think these numbers are?*
George: *It's 5 times...each time out of 50.*
Anna: *We have 28 the first time, 23 the second time, 24 the third time, etc.*
Instr.: *What does each of these outcomes stand for? What does, for example, 28 stand for?*
Anna: *28 heads...*
Instr.: *Out of how many trials?*
George: *Out of 50...50 coin tosses.*

Anna and George then collected the count of heads for a large number of samples and drew the resulting distribution. Specifically, they asked the software to repeat the experiment 1000 times. While Fathom© was repeating the experiment, they would observe the table "Measures Table" eventually fill with statistics and the histogram of these statistics approaching the normal distribution. Looking at the histogram of the resulting distribution (Figure 6), the two students made observations such as, for example, that 30 heads in 50 tosses was pretty likely, but that a number as high as 40 or higher is "rare since all numbers in the graph are smaller than 40". They also noted that the distribution of the numbers they collected looked "like bell shaped."

The "Coin Toss" task is a representative case of students utilizing the features of Fathom© to build their informal understanding of fundamental concepts of statistical inference including the central limit theorem, sampling distributions, confidence intervals and hypothesis testing. Fathom's structure encouraged Anna and George to appreciate the different levels of simulation involved when building a sampling distribution (Table 2; adapted from Erickson, 2001). The building activ-

Figure 6. Case table and histogram of sample statistics for the "coin toss" task: a measures collection of 1000 samples

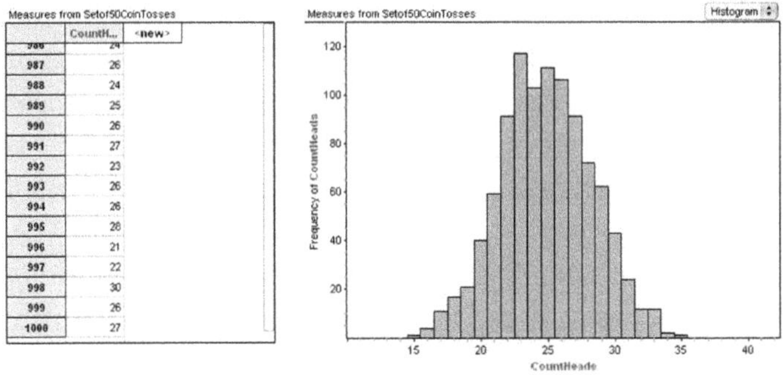

Table 2. Summary table of math SAT scores and FYGPA

Coin Toss Sample Collection	Measures collection
Each case is a single coin toss	Each case represents a collection of 50 coin tosses
Whole collection is a collection of 50 single coin tosses	The collection summarizes many samples of 50 coin tosses (5 by default)
The measure is a single number (a statistic) that describes the number of heads in the collection	Each case contains the count of heads of one collection of 50 coin tosses, so the collection has many counts of heads
Cannot estimate the mean count of heads here because this collection is only one sample	Can calculate the mean count of heads in this collection- by averaging the set of counts

ity of going from a Sample Collection to a Measures Collection helped them to begin to recognize the similarities and differences between population, sample, and sampling distribution.

The study participants engaged in several other activities similar in nature and format to the "Coin Toss" task. Students´ engagement with these activities resulted in them becoming very familiar both with the features of the software and the process of informally building the sampling distribution of different statistics and drawing inferences about the likely intervals of the corresponding population parameters. Using Fathom©, students built and modified their own simulations, and this allowed them, by working at their own pace, to make direct connections between the formal and the informal. When approaching the different tasks, they first had to simulate a single sample from some population they specified in the Formula Editor. Then, in the Measures Collection, they had to specify the statistic relevant for the problem. They saw how the Measures Collection calculated this statistic and turned it into a Case Attribute, which they would then graph and investigate its behaviour. Next, they would build the distribution of the statistic by repeating the process many times, collecting the statistic for each sample.

Since Fathom© is a dynamic system, students observed how the distribution of the statistic would change as the number of samples increased and would approach the sampling distribution of the statistic. Unlike standard simulation approaches to statistical inference, where there is a tendency to focus on relative frequencies or on means and to ignore variation, students were encouraged to explore the entire distribution of the statistic. They created informal confidence intervals and decided what range of values of the corresponding parameter they would consider reasonable, and which values seemed implausible or, in students' terminology, which were "rare events."

Grasping the ideas of sampling distributions and hypothesis testing, even at an informal level, was not easy. In the beginning, students had difficulties modelling the problems they were investigating using Fathom©. Also, they tended to confuse single samples with the samples of statistics obtained with the Measures Collection. However, unlike "black-box" simulations, where the user simply observes the computer build the sampling distribution, the structure of Fathom© and, especially the Measures Collection, helped learners understand the relationship between population, sample, and sampling distribution. This was a key breakthrough that helped build connections to inferential statistics. Understanding how population, sample, and sampling distribution are related is something that most university students taking introductory statistics courses fail to do. Although they may be able to calculate a standard deviation and a standard error, they "do not understand how these concepts are related (and distinguished) and so make application mistakes such as using one concept when they should use the other" (Schau & Mattern, 1997, p. 91).

Eventually, processes became more formal. The formal procedures for conducting hypothesis testing were introduced, and students began comparing empirical probabilities with the theoretical ones. Introduction of abstract notation caused a lot of difficulties to several study participants. Whereas the whole class seemed very comfortable with the process of informally building sampling distributions and making inferences, students got very puzzled and intimidated when formal notation and procedures were introduced. In the beginning, students had difficulties in using traditional tools such as z, and t. Eventually, however, the simulation approach gave them insights that helped them better understand and appreciate the meaning and power of these formal inferential tools. Of course, there were some students in the class who, due a poor mathematical background, even at the end of the semester still confused basic statistical notation. However, the conceptual understanding of sampling distributions and the logic of statistical inference that even these students had was much better compared to what is typically observed in students completing introductory statistics courses.

The "Is the New Drug Effective?" task in Figure 7 is an example of the type of activities that students had at the later stage of the course.

Figure 7. "Is the new drug effective?" task

FDA is the gatekeeper in the US to ensure that new drugs are thoroughly tested and proven effective before being made available to patients. Companies have to submit data to demonstrate that their new drug indeed significantly improves patients' condition. This usually involves many years of experiments on animals as well as on patients. Here, we are presenting a very simplified example of a new drug developed to reduce cholesterol level.
Company A submitted the following data and claimed that the new drug is very effective in reducing cholesterol level:
A new drug by Company A was given to a randomly selected sample of 64 patients whose cholesterol levels were at the high risk level of more than 250 mg. Before receiving the new drug, the mean cholesterol level of these 64 patients was 285 mg. The new drug was given to patients for a month. Their cholesterol level was measured again. The 64 cholesterol levels were summarized into Sample Mean = 250 mg, and Sample S.D. = 100 mg.
You work for the FDA and your job is to find out whether the claim made by Company A is valid or not.

"Is the New Drug Effective?" Task

In this group activity (See Figure 7), students had to decide whether the reduction in the average cholesterol level of a sample of 64 high-risk patients from 285 mg (Sample S.D. = 100 mg) to 250 mg was significant enough for the Federal Drug Administration (FDA) to consider the drug effective in reducing cholesterol level. A description of how a group of three students – Ioanna, Andreas, and Jennifer – worked in this task follows.

The students' first action was to open a new Fathom© file and then drag and drop an empty Case Table. Next, they started discussing about how to model the situation described in the first part of the task. As seen in the following discussion, the need to properly model the situation, forced students to think carefully about the problem and its context:

Instr.: *What is this problem about?*

Andreas: *About the drug. You want to decide whether the lowering of a cholesterol level that they found in the sample of 64 is a success of the drug. So, we should try to find out whether 250 was rare or not.*

Instr.: *Rare compared to what?*

Andreas: *Rare compared to not rare. (laughs)*

Instr.: *Rare compared to what? (laughs)*

Ioanna: *It's gonna go into this side, the lower side. We want to see if the level has gone down.*

Andreas: *So, you want to see if it's rare compared to 285, which was the mean level before. So you assume that 285 is…*

Ioanna: *the mean.*

Andreas: *And so check if the sample mean of 250 is rare given a population mean of 285.*

Instr.: *How are going to decide whether 250 is rare or not?*

Andreas: *You see if it's in the range or not.*

Jennifer: *Or if…There are just very few values that low.*

Instr.: *So, what is the null hypothesis?*

Ioanna: *That the drug does not work.*

Instr.: *So, if the drug does not work, what would be the mean cholesterol level of the patients?*

Ioanna: *285 and you want to see if 250 is rare given a mean of 285.*

Andreas: *So, the null is 285 and the alternative is to prove that it has improved.*

Instr.: *Not to prove…you can never prove…you're trying to see if there is enough evidence to reject the null hypothesis.*

Andreas: *That's right. You can never prove... So Ha is less than 285...?*
Instr.: *Great.*
Jennifer: *Now we have to get...we have to give it the formula for simulating the sample.*
Andreas: *What kind of distribution are we going to simulate the observations from?*
Anna: *I don't know.*
Andreas: *You will simulate from the null....Mean 285. Standard deviation 100.*
Anna: *And let's assume that the population is normal.*
Jennifer: *(puts down the formula) OK. Yeah. I'm gonna draw the histogram so that we can see.*
Andreas: *It seems quite spread out.*
Jennifer: *Because it's a 100 [the population standard deviation].*
Andreas: *What are you going to do now?*

Next, the three students used Fathom© to build the distribution of mean cholesterol levels for a large number of samples of size 64 coming from the population distribution they defined, which had a mean 285 and a standard deviation of 100. They put the statistics they collected on a histogram and noted that distribution of sample means was normally distributed and had a much lower spread than the distribution of the single sample they had first simulated. They also noted that 250 seemed rare, since "there are just very few values that low". In order to verify that one would reject the null hypothesis at the 95% confidence level, students drew a vertical line where the 5th percentile was located and saw that 250 was below the line (See Figure 8). They concluded that "the lowering of a cholesterol level is probably a success of the drug". They concluded that "the drug is effective since the level has gone down".

Students then compared the distribution of sample mean they got using the computer with the theoretical distribution of the mean (i.e. the sampling distribution) for samples of size 64 coming from a population with mean 285 and standard deviation 100. They drew the sampling distribution curve and saw that values were very close to the ones obtained by the computer: "One standard deviation below the mean is 272.5, 2 standard deviations below is 260…250 is further to the left." They saw that values were very close to the ones they got from the computer, "because one standard deviation is 12.5, two standard deviations is 25". They found the Z-score corresponding to 250, and saw that it was "about 2.8 standard deviations below the mean, this makes it a rare event."

The excerpts previously illustrate how for these students, who had been exposed to inferential statistics through an experimental approach that employed the features of the pedagogical software Fathom, the logic of inference became quite under-

Figure 8. Histogram of sample statistics for the "drug for reducing cholesterol level" task: a measures collection of 1000 samples

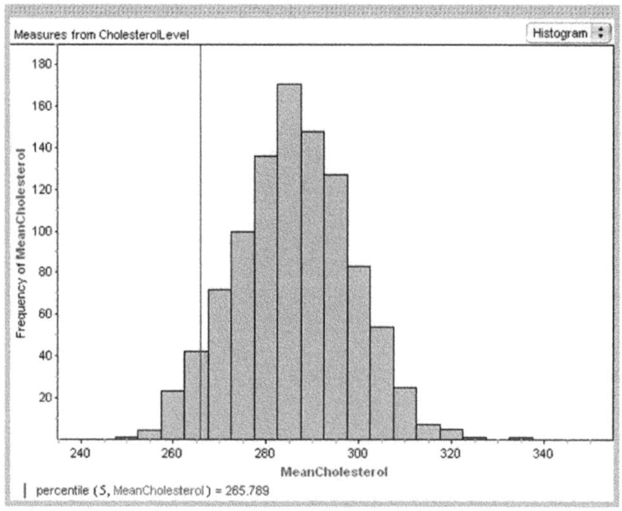

standable. In solving statistical problems, they were able to use both the empirical approach and the formal methods of inference they had learned in their statistics course. Although they still had difficulties in using traditional tools such as *z* and *t*, the simulation approach gave them insights that helped them better understand and appreciate the meaning and power of these formal inferential tools.

Lessons Learned from the Course Delivery

Traditionally, introductory statistics course adopt a linear, hierarchical approach to the different statistical topics encountered in the course, with inferential statistics being introduced in the later part of the course. It is assumed that this simplifies the process of learning by gradually leading students from more basic to more complex connections (Steinbring, 1990). However, presenting statistical content as a sequenced list of curricular topics might lead to compartmentalization of knowledge and fail to communicate to students the interconnectedness of the different statistical ideas they encounter in the course. The "progressive formalization" approach adopted in the current study, where instruction focuses on building students' conceptual understanding of the big picture of statistics by emphasizing and revisiting a set of central statistical ideas proved to be a promising alternative to the more conventional approach.

Analysis of the data collected during the instructional experiment, indicates that students' genuine statistical endeavors in a carefully designed, technology-based learning environment focusing on the "big ideas" of statistics, fostered students' ability to reason and make appropriate data-based inferences. Through their continuous participation in a variety of interesting, hands-on, and computer-supported activities that elicited conceptions of variability and difference rather than center and sameness, students "discovered the richness and complexity of data" (Hammerman & Rubin, 2003). The informal, data-driven inferences on which instruction focused on the first part of the course, helped students develop understandings of fundamental aspects of informal inference and argumentative reasoning (Ben-Zvi, 2006) that served as foundations for the formal study of inferential statistics in the latter part of the course.

The affordances offered by the dynamic software Fathom© for delving deeply into the data to make sense of the situation at hand, were instrumental in supporting student understanding of both informal and formal inferential statistics. Use of Fathom© supported active knowledge construction and extended students' ability to produce reasonable inferences and arguments. All the students became fully engaged in the genuine data explorations afforded through use of the dynamic statistics software. Through analyzing real data and performing computer-based simulations, they experimented with statistical ideas, made conjectures, and tested and refined these conjectures. The graphical and other data representations offered by technology, gave a focal point for discussions that helped students reflect on the validity of their data-based arguments. Students made connections between data and chance, and thus improved their understanding of the bidirectional relationship between empirical data distributions and theoretical probability distributions (Stohl & Tarr, 2002).

Fathom© structure encouraged students to experiment with statistical ideas, to make conjectures and use the investigative tools provided by the software to test and modify these conjectures. Use of such software in the classroom supported active knowledge construction by "doing" and "seeing" statistics in a powerful and flexible learning environment (Ben-Zvi, 2000). Features like the "drag and drop" interface, i.e. the fact that students have to drag rather than choose things, increased immediacy of data analysis (Pratt, 1998). Attributes like the ability to dynamically link multiple representations, to provide immediate feedback, and to transform a whole representation into a manipulable object, provided affordances towards more constructive pedagogical approaches than the attributes of professional statistics software. Use of the Formula Editor helped formal ideas become more concrete by providing students with a control mechanism that gave them the opportunity to "express fuzzy ideas in a formal, conventional and rigorous language" (Pratt, 1998, p. 108). At the same time, formulas resembled everyday language enough to be easy to learn, making Fathom© a user-friendly environment.

Concurring with findings of previous studies conducted by the author (e.g. Meletiou-Mavrotheris, 2003; Meletiou-Mavrotheris & Lee, 2007), the unique features of the dynamic software proved particularly effective in supporting student understanding of the fundamental concept of sampling distribution, a stumbling stone to formal statistical inference. Through investigation of the ways in which students used Fathom© to construct the sampling distribution of different statistics and, based on the resulting distribution, to draw inferences regarding the corresponding population parameter, the following student behaviors, which might be more difficult to achieve in a more traditional statistics classroom were observed: (1) students were actively engaged in constructing their own understanding of sampling distribution; (2) students were able to verbally explain how a sampling distribution is constructed, and how it can be used to draw informal inferences as to whether an event is "rare" or "not rare"; (3) students were able to make multiple presentations of sampling distributions, both graphically and numerically; and (4) students were able to provide proper explanations of the difference between population distribution, sample data distribution and sampling distribution of a statistic. Although, due to the exploratory nature of the study design, these observations were not further assessed or compared with other more traditional classroom formats, they do suggest that integration of Fathom© into instruction brought about important changes in students' ways of learning. Students got to appreciate "the different layers of simulations" involved when building the distribution of a statistic (Erickson, 2001), and this helped them construct a fairly sophisticated understanding of both sampling distributions and the logic of inferential statistics.

FUTURE RESEARCH DIRECTIONS

A drawback of this case study is the limited generalizability of its findings. The qualitative methodology used to research the case, the small scale of the study, and its limited geographical nature, means that generalizations to cases that are not very similar should be done cautiously as the specific classroom investigated might not be representative of all introductory statistics classrooms. Additionally, although the rich insights gained from the study suggest that Fathom© holds a lot of promise as a tool for building student's both informal and formal inferential reasoning, the research design of the present inquiry does not allow the drawing of generalizations.

Future effect studies of the software, using an experimental research design, should be carried out to assess the potential of the software as a tool for supporting the development of introductory statistics students' inferential reasoning. Affective factors associated with the use of dynamic software in an active learning environment, and the impact these might have on student learning is another area that ought to be investigated in future research.

CONCLUSION

Although technology has for a while been employed in statistics instruction, the packages traditionally used are not pedagogical tools, but professional statistical software, or simplified versions of such software, designed primarily for statistical analysis rather than instruction. This case study was designed to investigate the potential of the statistical package Fathom©, an educational software specifically developed for statistics learning, as an aid to the development of college-level students' conceptions of statistical inference.

Findings from the current case study, although not generalizable to other settings given the nature of the study, still provide useful information about technology use in introductory statistics classrooms. They point to a number of critical ingredients that statistics educators should consider when choosing statistics software, including the following; (1) a dynamic nature which permits the construction and flexible usage of multiple data representations; (2) a student-friendly interface that facilitates learning rather than add to the mental constraints that students have to overcome; (3) access to good visualizations that help students expand their understanding of difficult statistical concepts (e.g. the powerful visualization tools provided by Fathom's Measures Collection that help students overcome the conceptual difficulties involved in understanding sampling distributions); (4) constructive representations that support content and pedagogy by affording learners with tools they can use to construct their own conceptual understanding.

The insights gained from the study indicate that introduction to fundamental notions of formal inferential statistics such as sampling distribution and confidence intervals through a technological tool such as Fathom©, which is specifically designed to encourage students to build, refine, and reorganize their prior understandings and intuitions about the stochastic seems to be promising alternative to instruction that adopts the more conventional black-box approach. Black-box simulations, where the user simply watches the computer build the sampling distribution do not allow direct connections between the formal and the informal. Dynamic software like Fathom©, on the other hand, provide a medium for the design of activities that integrate experiential and formal pieces of knowledge and allow the user to connect the symbolic and the iconic (Pratt, 1998). The interaction between the data and the theoretical model seems to be more convincing than black-box simulations, and to help students construct more powerful meanings.

The advantages of Fathom© stem from the fact that, in contrast to software based on black-box simulations where the implicit conjecture of the software design often is that students will recognize the key assumptions underlying the computer output, Fathom's design involved developing conjectures based on a conceptual (or epistemological) analysis of key statistical ideas (e.g. population, sample, sampling distribution). This conceptual analysis led to the development of a package with an intuitive interface and a structure that allows students to play an active role in the construction of fundamental statistical ideas. For example, the idea of *Measures Collection* makes Fathom© quite unique with respect to informally introducing the notion of sampling distributions.

The conceptual difficulties students have to overcome when working with experimental probabilities using a software such as Fathom© are similar to those they encounter when dealing with theoretical probabilities. They still have to understand the real problem to model it properly. However, the experimental approach gives students a more concrete take on the problems. Of course, since simulations take longer and empirical probabilities do not exactly match their theoretical counterparts, it is still useful to learn the streamlined statistics of *t* tables and professional packages (Erickson, 2001). Nonetheless, because the logic of inference is so much more understandable through the simulation approach used by a dynamic tool such as Fathom©, the experimental approach should work hand-in-hand with the more formal approach.

REFERENCES

Bakker, A. (2003). Reasoning about shape as a pattern in variability. In Lee, C. (Ed.), *Reasoning about Variability: A Collection of Current Research Studies*. Dordrecht, The Netherlands: Kluwer Academic Publisher.

Bakker, A. (2004). *Design research in statistics education: On symbolizing and computer tools*. (Doctoral dissertation). Utrecht University, Utrecht, The Netherlands.

Ben-Zvi, D. (2000). Toward understanding the role of technological tools in statistical learning. *Mathematical Thinking and Learning*, *2*, 127–155. doi:10.1207/S15327833MTL0202_6.

Ben-Zvi, D. (2006). Scaffolding students' informal inference and argumentation. In A. Rossman & B. Chance (Eds.), *Working Cooperatively in Statistics Education: Proceedings of the Seventh International Conference of Teaching Statistics*. Salvador, Brazil: ICOTS.

Biehler, R. (1995). *Toward requirements for more adequate software tools that support both: Learning and doing statistics*. Bielefeld, Germany: University of Bielefeld.

Erickson, T. (2001). *Data in depth: Exploring mathematics with fathom*. Emeryville, CA: Key Curriculum Press.

Erickson, T. (2006). Using simulation to learn about inference. In A. Rossman & B. Chance (Eds.), *Working Cooperatively in Statistics Education: Proceedings of the Seventh International Conference of Teaching Statistics* (ICOTS-7). Salvador, Brazil: ICOTS.

Finzer, W. (2000). *Design of fathom: A dynamic statistics environment, for the teaching of mathematics*. Retrieved July 2, 2011 from http://www.keypress.com/Documents/fathom/Bibliography/Fathom_in_Teaching_Math.pdf

Finzer, W. (2001). *Fathom*. Emeryville, CA: Key Curriculum Press.

Finzer, W., & Erickson, T. (1998). DataSpace — A computer learning environment for data analysis and statistics based on dynamic dragging, visualization, simulation, and networked collaboration. In *Proceedings of the Fifth International Conference on Teaching of Statistics* (pp. 825-830). Voorburg, The Netherlands: International Statistical Institute.

Franklin, C. A., Kader, G., Mewborn, D., Moreno, J., Peck, R., Perry, M., & Scheaffer, R. (2007). *Guidelines for assessment and instruction in statistics education (GAISE) report: A pre-K–12 curriculum framework*. Alexandria, VA: American Statistical Association.

Gardner, H., & Hudson, I. (1999). University students' ability to apply statistical procedures. *Journal of Statistics Education, 7*(1). Retrieved July 2, 2011 from www.amstat.org/publications/jse/secure/v7n1/gardner.cfm

Garfield, J., & Ahlgren, A. (1998). Difficulties in learning basic concepts in probability and statistics: Implications for research. *Journal for Research in Mathematics Education, 19*, 44–63. doi:10.2307/749110.

Gil, E., & Ben-Zvi, D. (2011). Explanations and context in the emergence of students' informal inferential reasoning. *Mathematical Thinking and Learning, 13*(1), 87–108. doi:10.1080/10986065.2011.538295.

Gordon, F. S., & Gordon, S. P. (1992). Sampling + simulation = statistical understanding. In Gordon, F. S. (Ed.), *Statistics for the Twenty-First Century* (pp. 207–216). Washington, DC: The Mathematical Association of America.

Green, D. G. (1982). A survey of probability concepts in 3000 students aged 11-16. In D. V. Grey (Ed.), *Proceedings of the First International Conference on Teaching Statistics* (pp. 766-783). London: Statistics Teaching Trust.

Hammerman, J., & Rubin, A. (2003). Reasoning in the presence of variability. In Lee, C. (Ed.), *Reasoning about Variability: A Collection of Current Research Studies*. Dordrecht, The Netherlands: Kluwer Academic Publisher.

Hoyles, C., & Noss, R. (1994). Dynamic geometry environments: What is the point? *Mathematics Teacher, 87*, 716–717.

Konold, C. (2002). Teaching concepts rather than conventions. *New England Journal of Mathematics, 34*(2), 69–81.

Konold, C., & Miller, C. D. (2005). *TinkerPlots: Dynamic data exploration*. Emeryville, CA: Key Curriculum Press.

Makar, K., & Rubin, A. (2007). Beyond the bar graph: Teaching informal statistical inference in primary school. In *Proceedings of the Fifth International Research Forum on Statistical Reasoning, Thinking, and Literacy (SRTL-5)*. Warwick, UK: SRTL.

Meletiou-Mavrotheris, M. (2003). Technological tools in the introductory statistics classroom: Effects on student understanding of inferential statistics. *International Journal of Computers for Mathematical Learning, 8*(3), 265–297. doi:10.1023/B:IJCO.0000021794.08422.65.

Meletiou-Mavrotheris, M., & Lee, C. (2007). Effects of τechnological tools on introductory statistics students' understanding of sampling distributions. In Murtonen, M., Rautopuro, J., & Väisänen, P. (Eds.), *Learning and Teaching of Research Methods at University* (pp. 95–118). Turku, Finland: University of Turku.

Meletiou-Mavrotheris, M., Paparistodemou, E., & Stylianou, D. (2009). Enhancing statistics instruction in elementary schools: Integrating technology in professional development. *The Montana Math Enthusiast, 6*(1&2), 57–78.

National Research Council. (2000). *How people learn: Brain, mind, experience, and school*. Washington, DC: National Academy Press.

Noss, R., & Hoyles, C. (1996). *Windows on mathematical meanings: Learning cultures and computers*. Dordrecht, The Netherlands: Kluwer Academic Publishers. doi:10.1007/978-94-009-1696-8.

Paparistodemou, E., & Meletiou-Mavrotheris, M. (2008). Developing young students' informal inference skills in data analysis. *Statistics Education Research Journal, 7*(2), 83–106.

Paparistodemou, E., & Meletiou-Mavrotheris, M. (2010). Engaging young children in informal statistical inference. In C. Reading (Ed.), *Data and Context in Statistics Education: Towards an Evidence-Based Society: Proceedings of the Eighth International Conference on Teaching Statistics* (ICOTS8). Ljubljana, Slovenia: International Statistical Institute.

Paparistodemou, E., & Noss, R. (2004). Designing for local and global meanings of randomness. In *Proceedings of the Twentieth Eighth Annual Conference of the International Group for the Psychology of Mathematics Education* (pp. 497-504). Bergen, Norway: PME.

Paparistodemou, E., Noss, R., & Pratt, D. (2008). The interplay between fairness and randomness in a spatial computer game. *International Journal of Computers for Mathematical Learning, 13*(2), 89–110. doi:10.1007/s10758-008-9132-8.

Pfannkuch, M. (2006). Informal inferential reasoning. In A. Rossman & B. Chance (Eds.), *Working Cooperatively in Statistics Education: Proceedings of the Seventh International Conference ofTeaching Statistics (ICOTS-7)*. Salvador, Brazil: ICOTS.

Pfannkuch, M., Budgett, S., Parsonage, R., & Horring, J. (2004). *Comparison of data plots: Building a pedagogical framework.* Paper presented at the International Association for Statistics Education 2004. Lund, Sweden: IASE.

Pratt, D. (2000). Making sense of the total of two dice. *Journal for Research in Mathematics Education, 31*, 602–625. doi:10.2307/749889.

Pratt, D., Johnston-Wilder, P., Ainley, J., & Mason, J. (2008). Local and global thinking in statistical inference. *Statistics Education Research Journal, 7*(2), 107–129.

Pratt, D. C. (1998). *The construction of meanings in and for a stochastic domain of abstraction.* (Ph.D. Thesis). University of London, London, UK.

Reichardt, C. S., & Gollob, H. F. (1997). When confidence intervals should be used instead of statistical tests, and vice versa. In Harlow, L. L., Mulaik, S. A., & Steiger, J. H. (Eds.), *What if There Were no Significance Tests?* (pp. 259–284). London: Lawrence Erlbaum Associates.

Rubin, A., Bruce, B., & Tenney, Y. (1990). *Learning about sampling: Trouble at the core of statistics.* Paper presented at the Annual Meeting of the American Educational Research Association. New York, NY.

Rubin, A., Hammerman, J., & Konold, C. (2006). Exploring informal inference with interactive visualization software. In A. Rossman & B. Chance (Eds.), *Working Cooperatively in Statistics Education: Proceedings of the Seventh International Conference of Teaching Statistics* (ICOTS-7). Salvador, Brazil: ICOTS.

Schau, C., & Mattern, N. (1997). Assessing students' understanding of statistical relationships. In Gal, I., & Garfield, J. B. (Eds.), *The Assessment Challenge in Statistics Education*. Burke, VA: IOS Press.

Shamatha, J. H., Peressini, D., & Meymaris, K. (2004). Technology-supported mathematics activities situated within an effective learning environment theoretical framework. *Contemporary Issues in Technology and Teacher Education, 3*(4). Retrieved July 2, 2011, from http://www.citejournal.org/vol3/iss4/mathematics/article1.cfm

Stake, R. (1995). *The art of case study research*. Thousand Oaks, CA: Sage Publications.

Steinbring. (1990). The use of chance-concept in everyday teaching - Aspects of a socially constituted epistemology of mathematical knowledge. In J. B. Garfield (Ed.), *Research Papers from the Third International Conference on Teaching Statistics.* Dunedin, New Zealand: University of Otago.

Stohl, H., & Tarr, J. E. (2002). Using multi-representational computer tools to make sense of inference. In D. Mewborn (Ed.), *Proceedings of the Twenty-Fourth Annual Meeting of the North American Chapter of the International Group for the Psychology of Mathematics Education*. Athens, GA: PME.

Stylianou, D., & Meletiou-Mavrotheris, M. (2005). Graphical representation of data: The effect of the use of a dynamical statistics technological tool. *Scientia Pedagogica Experimentalis, 42*(2), 273–294.

Watson, J. M. (2008). Exploring beginning inference with novice grade 7 students. *Statistics Education Research Journal, 7*(2), 59–82.

Wilensky, U. (1993). *Connected mathematics - Building concrete relationships with mathematical knowledge.* (Ph.D. Thesis). Massachusetts Institute of Technology, Cambridge, MA.

Yu, C. H., & Behrens, J. T. (1994). Identification of misconceptions in learning statistical power with dynamic graphics as remedial tool. *ASAProStEd*, 242-246.

Zieffler, A., Garfield, J., delMas, R., & Reading, C. (2008). A framework to support research on informal inferential reasoning. *Statistics Education Research Journal, 7*(2), 40–58.

ADDITIONAL READING

Bakker, A. (2002). Route-type and landscape-type software for learning statistical data analysis. In B. Phillips (Ed.), *Proceedings of the Sixth International Conference on Teaching Statistics* (ICOTS-6). Cape Town, South Africa: International Statistical Institute.

Biehler, R. (2003). Interrelated learning and working environments for supporting the use of computer tools in introductory classes. In *Proceedings of IASE Satellite Conference on Statistics Education and the Internet.* Voorburg, The Netherlands: International Statistical Institute.

Canal, G. Y. (2002). Some challenges for the use of computer simulation for solving conditional probability problems. In D.S. Mewborn, P. Sztajn, D.Y. White, H.G. Weigel, R.L. Bryant, & K. Nooney (Eds.), *Proceedings of the Twenty-Fourth Annual Meeting of the North American Chapter of the International Group for the Psychology of Mathematics Education* (pp. 1255-1266). Columbus, OH: ERIC Clearinghouse for Science, Mathematics, and Environmental Education.

Carver, R. (2001). Fathom dynamic statistics software. *The Statistics Teacher Network, 56*, 1–4.

Collings, P. B. (2001). Fathom: Dynamic systems software [review]. *The American Statistician, 55*(3), 258–263. doi:10.1198/tas.2001.s123.

Confrey, J., & Makar, K. (2005). Critiquing and improving data use from high stakes tests: Understanding variation and distribution in relation to equity using dynamic statistics software. In Dede, C., Honan, J. P., & Peters, L. C. (Eds.), *Scaling Up Success: Lessons Learned from Technology-Based Educational Improvement* (pp. 198–226). San Francisco, CA: Jossey-Bass.

Cradler, J., McNabb, M., Freeman, M., & Burchett, R. (2002). How does technology influence student learning? *Learning and Leading with Technology, 29*(8), 46–56.

Erickson, T. E., & Finzer, W. F. (1998). What to teach before inference: Building the bones from paper, software, and stories. In *Proceedings of the Fifth International Conference on Teaching of Statistics.* Voorburg, The Netherlands: International Statistical Institute.

Feldman, A., Konold, C., & Couter, B. (2000). *Network science a decade later: The internet and classroom learning*. Mahwah, NJ: Lawrence Erlbaum Associates Inc..

Hesterberg, T. (2006). Bootstrapping students' understanding of statistical concepts. In Burrill, G. F. (Ed.), *Sixty-Eighth Yearbook: Thinking and Reasoning with Data and Chance* (pp. 391–416). Reston, VA: NCTM.

Jonassen, D., Howland, J., Marra, R. M., & Crismond, D. (2008). *Meaningful learning with technology*. Upper Saddle River, NJ: Pearson Education Inc..

Kazak, S. (2006). *Investigating elementary school students' reasoning about distributions in various chance events.* (Unpublished doctoral dissertation). Washington University, St. Louis, MO.

Konold, C. (2007). Designing a data tool for learners. In M. Lovett & P. Shah (Eds.), *Thinking with Data: The 33rd Annual Carnegie Symposium on Cognition* (pp. 267-292). Hillside, NJ: Lawrence Erlbaum Associates.

Konold, C., & Kazak, S. (2008). Reconnecting data and chance. *Technology Innovations in Statistics Education, 2*(1). Retrieved July 2, 2011, from http://escholarship.org/uc/item/38p7c94v

Konold, C., & Lehrer, R. (2008). Technology and mathematics education: An essay in honor of Jim Kaput. In English, L. D. (Ed.), *Handbook of International Research in Mathematics Education* (2nd ed., pp. 49–72). New York: Routledge.

Lane-Getaz, S. J. (2006). What is statistical thinking and how is it developed? In Burrill, G. F., & Elliott, P. C. (Eds.), *Sixty-eighth yearbook: Thinking and reasoning with data and chance* (*Vol. 68*, pp. 273–290). Reston, VA: NCTM.

Lipson, K. (2002). The role of computer based technology in developing understanding of the concept of sampling distribution. In B. Phillips (Ed.), *Proceedings of the Sixth International Conference on Teaching of Statistics* (ICOTS-6). Cape Town, South Africa: International Statistical Institute.

Lock, R. H. (2002). Using fathom to promote interactive explorations of statistical concepts. In B. Phillips (Ed.), *Proceedings of the Sixth International Conference on Teaching of Statistics* (ICOTS-6). Cape Town, South Africa: International Statistical Institute.

Meletiou-Mavrotheris, M., Lee, C., & Fouladi, R. (2007). Introductory statistics, college student attitudes and knowledge – A qualitative analysis of the impact of technology-based instruction. *International Journal of Mathematical Education in Science and Technology, 38*(1), 65–83. doi:10.1080/00207390601002765.

Mills, J. D. (2002). Using computer simulation methods to teach statistics: A review of the literature. *Journal of Statistics Education, 10*(1).

Saldanha, L. A., & Thompson, P. W. (2003). Conceptions of sample and their relationship to statistical inference. *Educational Studies in Mathematics*, *51*, 257–270. doi:10.1023/A:1023692604014.

Sánchez, E. (2006). Meanings' construction about sampling distributions in a dynamic statistics environment. In A. Rossman & B. Chance (Eds.), *Working Cooperatively in Statistics Education: Proceedings of the Seventh International Conference of Teaching Statistics* (ICOTS-7). Salvador, Brazil: ICOTS.

Shafer, K. (2010). Scrambling data with fathom to simulate the null hypothesis. *Mathematics Teacher*, *103*(6), 453–457.

van der Meij, J., & de Jong, T. (2006). Supporting students' learning with multiple representations in a dynamic simulation-based learning environment. *Learning and Instruction*, *16*(3), 199–212. doi:10.1016/j.learninstruc.2006.03.007.

Zbiek, R. M., Heid, M. K., Blume, G. W., & Dick, T. P. (2007). Research on technology in mathematics education. In Lester, F. K. (Ed.), *Second Handbook of Research on Mathematics Teaching and Learning* (*Vol. 2*, pp. 1169–1208). Charlotte, NC: Information Age Publishing.

Section 3
Outside the Norm

Chapter 11
Coping with Infinity:
Using TI-*N*spire™ CAS to Bring Alive Multiple Representations in Mathematics

Bjørn Felsager
Midtsjællands Gymnasieskoler, Denmark

ABSTRACT

This chapter outlines the structure of the initial module of a pilot project investigating the benefits of using dynamical mathematical software when teaching students about cross-curricular models of thinking. Focus is on the use of conceptual metaphors when teaching the concept of infinity in mathematics as well as in literature. The use of dynamical mathematics software facilitated constructions of multiple representations in problem solving in a series of carefully selected problems involving infinity. The pilot project is currently being evaluated, but preliminary findings indicate an increased student awareness of the usefulness of the software when learning how to use mathematical inscriptions for problem solving. The pilot project will be expanded in the Fall 2011 to include 6 classes across Denmark to test the applicability of the software, the teaching model and the teaching materials in a variety of classes to see if the initial successes of the project can be replicated.

DOI: 10.4018/978-1-4666-4050-4.ch011

INTRODUCTION

In a recent project concerning "Mathematics and Cognition" we introduced a pre-calculus class to the dynamical mathematical software TI-*N*spire™ CAS—from Texas Instruments (2007)—and confronted them with the challenges of understanding infinity in a mathematical context. The purpose of the project was three-fold:

1. To introduce the class to the use of multiple representations using various mathematical inscriptions (graphs, tables, equations, text …) in problem solving.
2. To introduce the class to the theory of modern cognition, especially the concepts of metaphor and blending with the purpose of making them aware of some of the basic cognitive mechanisms underlying abstract thought. This would also be useful for the students in humanistic subjects (literature).
3. To introduce the class to the fundamental ambiguous concept of infinity the taming of which is a central part of the history of mathematics.

In this chapter, I will focus exclusively upon the mathematical didactical aspects of the project. The project was structured in several parts, where I will only go into details concerning the first part. The first part was subdivided into a general introduction, several workshops, and a concluding session, which incorporated a summary of what they have learned so far.

The introductory part consisted of a general introduction to mathematics and cognition with the emphasis upon conceptual metaphors and blendings and their role in facilitating communication in mathematics.

The workshop part consisted of the students working on various problems using multiple presentations, including a number of problems concerned with infinity.

The concluding session was devoted to a cross curricular investigation of the novel "The Book of Sand" written by Borges (1975).

General Considerations Preceding the Project

To make it possible to work with multiple representations we needed a versatile and user friendly dynamical mathematical software tool. In our school, we have adopted the computer program TI-*N*spire™ CAS from Texas Instruments, which is an integrated computer package specifically designed to facilitate the use of multiple representations in mathematics. The program consists of a number of dynamically interactive applications among other subjects offering support for investigations in geometry, statistics, tables, graphing as well as symbolic calculations (hence the

CAS affixed to the name TI-*N*spire™). The class has mathematics as an integrated subject and consisted of three kinds of students; some having only mathematics for one year (C-level), some having mathematics for two years (B-level) and some having mathematics for three years (A-level). As prerequisite for the project all students were familiar with elementary geometry and in particular similarities, as well as a basic understating of linear relationships between numerical variables. Roughly speaking this comprises the beginning of the precalculus level.

Since calculus was ruled out we decided to take as a starting point Zeno's paradoxes (later in the project we also introduced vanishing points in perspective drawings and infinitely distant points attached to a Euclidean plane, but that is beyond the scope of the first part of the project I am dealing with in this paper).

Content of the Lessons

The mathematical section of the introductory part consisted in a presentation given by myself to conceptual metaphors and blendings based upon the so called "Riddle of the Buddhist monk" from Fauconnier and Turner (2002) – for a more recent discussion see also the Website Turner (2010). We started presenting the students with an animation of the riddle totally constructed in the TI-*N*spire™ CAS program and asked for the student's intuition about the solution of the riddle. Not surprisingly opinions differed among the students, but once they saw the actual visualization of the blending they got the expected AHA-experience as well as a valuable prototype of a conceptual blending. Then we moved on to incorporate basic metaphors involving "Time as Space" and discussed the inscription of graphs with time on the first axis, the path of the monk along the second axis and generalized the riddle to a general abstract mathematical setting, where the solution of the riddle is equivalent to a fixed point theorem for continuous functions (See Figures 1 and 2).

Next the students worked on the medieval problem of "The dog chasing the hare" from the collection of Alcuin as translated by Hadley and Singmaster (1992). The problem was presented for them in the original Latin (where they were asked to identify roman numerals and some keywords) as well as a modern Danish translation:

Problem 26[1] - The Dog Chasing the Hare

There is a field 150 feet long. At one end is a dog, and at the other a hare. The dog chases when the hare runs. The dog travels 9 feet in a jump, while the hare travels 7 feet. How many feet will be travelled by the pursuing dog and the fleeing hare before the hare is seized?

Figure 1. The riddle of the Buddhist monk (animation). The animation helps making the riddle come alive.

Figure 2. The blending of the first and third day (animation). The dynamical linking to the graph also underlies the important role of the important metaphor "time as space."

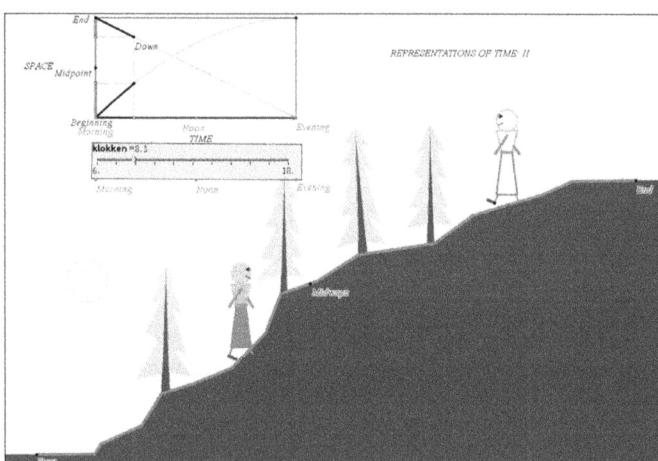

The students were informed of the lack of mathematical inscriptions in the problem, which only consisted of text and numbers, e.g. there were no diagrams explaining the problem. It is also notable that explicit mention of time is missing being replaced by "the number of jumps". Thus, there is an implicit hidden assumption, that the jump of the dog and the jump of the hare last the same time-interval, an assumption that must be part of a well-known medieval framework not worth mentioning to the reader.

The students were asked to transfer the information to a spreadsheet and associate graphs with their variables, which allowed them to solve the problem both numerically and graphically (See Figures 3 and 4).

Figure 3. Table and graph solution of "dog chasing a hare." Students actively construct and manipulate the tables and graphs. Without a dynamic software this would not be possible.

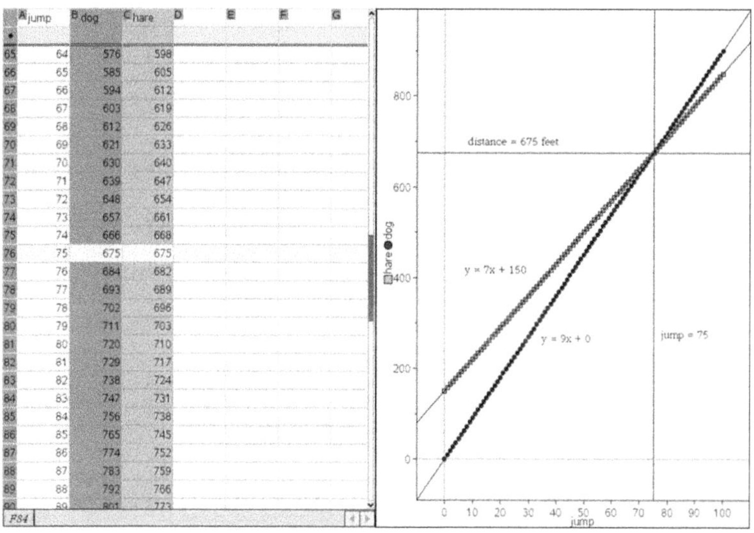

Figure 4. Symbolic solution of "dog chasing a hare." Many students are very poor in algebra, so the symbolic assistance of the program when students manipulate the equations is invaluable.

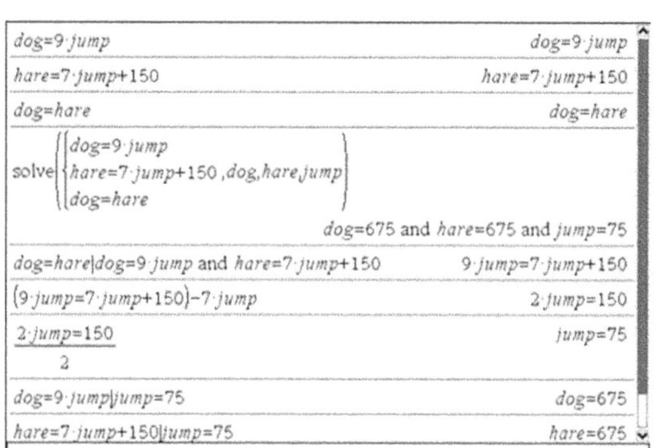

They also obtained the equations for the graphs using linear regression and interpreted the coefficients of the equations. And finally they switched to a symbolic environment, where they could solve the equations both with and without using the solve-command. Coming thus far they were shown the original solution (with lots of implicit hidden assumptions) and discussed the meaning of the solution in relationship to their own solution using multiple representations.

Solution

The length of the field is 150 feet. Take half of 150, which is 75. The dog goes 9 feet in a jump. 75 times 9 is 675; this is the number of feet the pursuing dog runs before he seizes the hare in his grasping teeth. Because in a jump the hare goes 7 feet, multiply 75 by 7, obtaining 525. This is the number of feet the fleeing hare travels before it is caught.

Thus empowered with both a working knowledge of the dynamical software and how to use it to create multiple representations, they were finally ready for coming to grips with the concept of infinity using the Riddle of Achilleus and the Tortoise as a starting point. This was reproduced graphically by the students, so that they got a grasp of how the infinite process is actually linked to the intersection of the two linear graphs representing "Achilleus and the Tortoise" (See Figure 5).

Zeno's Paradox: Achilleus and the Tortoise

In a race, the quickest runner can never overtake the slowest, since the pursuer must first reach the point whence the pursued started, so that the slower must always hold a lead. -Aristotle, Physics VI:9, 239b15

Subsequently several variants of Zenos paradoxes were presented to them such as the fly moving forth and back between two approaching locomotives. In order to strengthen the paradoxical nature of the argument involving (1) an infinite process the outcome of which is outside the scope of the infinite process and (2) that there is no direct link between what occurs within the infinite process and what happens once the infinite process has been completed after a finite time we next introduced them to The Riddle of Thomsons lamp and the concept of super tasks, Thomson (1954).

A lamp is switched on and off at the end of time intervals, the first of which is 1 second, the next ½ second, the next ¼ second, etc. For how long does the lamp switch on and off and what is the state of the lamp once we are past the switching on and off? (See Figures 6 and 7).

Figure 5. Achilleus and the tortoise: a dynamical model constructed by the students with sliders representing the speed ratio a and the initial lead b

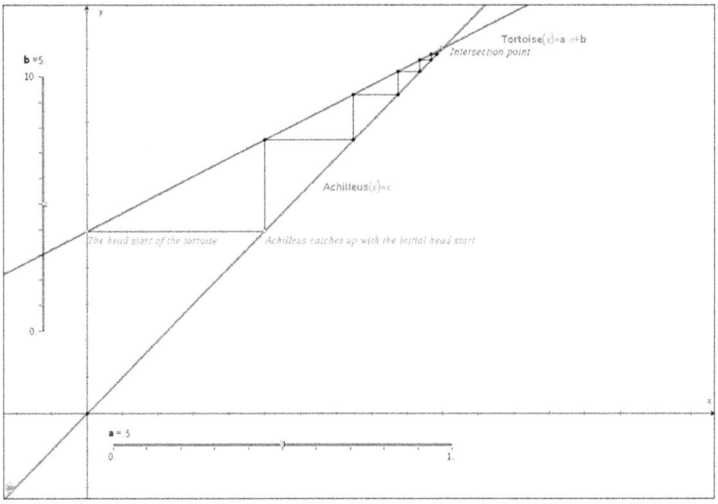

Figure 6. Table representation of the status of the Thompson lamp at various time intervals

	A interval	B time	C status	D value	E	F	G	H
1	1	0.	ON	1				
2	1	1.	ON	1				
3	2	1.	OFF	-1				
4	2	1.5	OFF	-1				
5	3	1.5	ON	1				
6	3	1.75	ON	1				
7	4	1.75	OFF	-1				
8	4	1.875	OFF	-1				
9	5	1.875	ON	1				
10	5	1.9375	ON	1				
11	6	1.9375	OFF	-1				
12	6	1.96875	OFF	-1				

Finally, they were ready to tackle a truly ambiguous problem in mathematics: The Riddle of the dog running back and forth between a boy and a girl happily marching along a road, Austin (1971) first published in the *Mathematics Magazine*.

A boy, a girl and a dog start at the same point facing in the same direction. The boy and girl start walking forward, the boy at 4 mph, the girl at 3 mph. The dog runs at 10 mph backwards and forwards between the boy and girl - when it meets one of them, it turns around and runs towards the other. After an hour, where is the dog and which way is it facing? (See Figure 8).

Figure 7. Graph representation of the status of the Thompson lamp with switch ON above time axis and switch OFF below

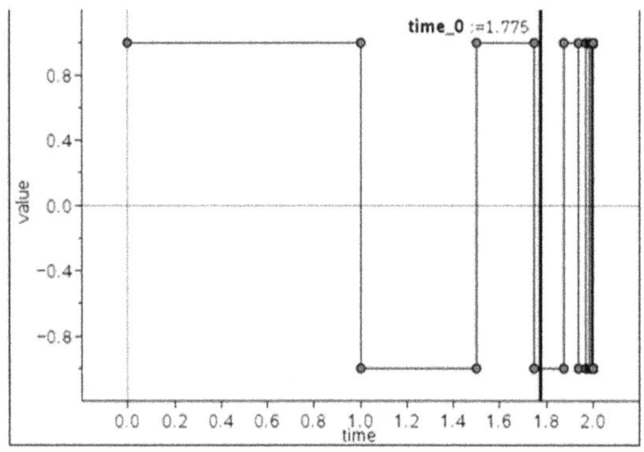

As shown, the problem can be presented to the students using an animation written entirely in TI-*N*spire™ CAS. But once we were certain the students understood the problem they tried to solve it themselves via a geometrical/graphical approach. Using the graphs of the boy and the girl as a starting point they selected a possible end position for the dog and a possible end direction, the dog is facing after 1 hour. They then reconstructed the path of the dog working backwards towards the origin with each segment of the dog's path being parallel to lines having slope

Figure 8. Animation of the dog running forth and back betweenn the boy and the girl. The students subseqently construct the graphs themselves.

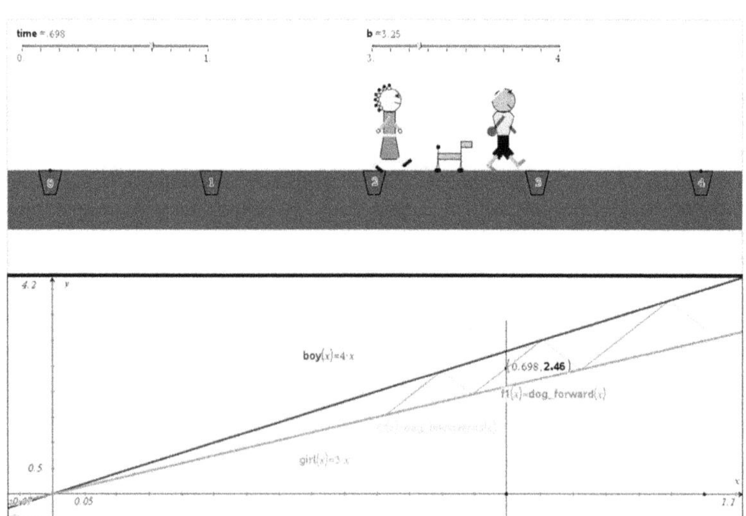

10 or -10. In this process, they discovered for themselves that the problem is truly ambiguous: Any position between the boy and the girl after 1 hour for the dog is compatible with the information stated in the problem. Similarly, any direction the dog is facing is possible at that time. But unlike Thomson's lamp this is not a problem of predicting the final state, but rather the time-inverted problem of not being able to predict the initial state of the dog. Both the "lamp"-problem and "the trotting dog"-problem however illustrated the inherent ambiguous nature of mathematics once infinite processes are included within the scope of problem solving.

When we summarized this first part of the project, we pointed out the role of the basic metaphor of infinity as stated in Lakoff and Núñez (2000). At this stage, we already had a number of cases at hand to illustrate the metaphor. The first part concluded with an investigation of the beautiful novel of Borges (1975), the Book of Sand, which the student read in a special hypertext edition on the Internet, Maximum Clarke (2001). In this short story of Borges the main character experiences what it would be like to be able to hold in your hand a truly infinite object, in this case a book with an infinite number of pages. This also made it possible to discuss various models of how to construct such a book depending upon the delicate structure of the infinite but countable number of pages. Especially it could be modeled using a traditional numbering of the countable set of rational numbers using a Farey type approach. This would result in a book having a dense set of pages in the sense that between any two pages, there would be an infinite number of pages, as well as the disorienting fact, that the book would contain no first and no last page.

CONCLUSION

This project is part of a bigger project concerning the role of cognition in the teaching of mathematics (as well as the teaching of literature). The "Coping with Infinity" project is currently being evaluated by the didactical researcher Peter Kaspersen (now emeritus) from Southern Danish University. Due to lack of funding it was not possible to record and analyze the individual lessons. However the project concluded with a series of presentations of mini projects made by the students. The projects were open ended projects where the students had to design a presentation using both mathematical and humanistic perspectives on a topic by their own choice that allowed them to combine conceptual metaphors and blendings as well as multiple representations with riddles concerning infinity. As examples of such topics, one group would take as a focal point the relationship between mirrors and infinity, another group would reenact the story of Hilberts Hotel, a third group would dramatize

Zenos paradoxes within the frame work of a classical fairytale with a woodsman never being able to return to his house and a dog exchanging an infinite number of messages between the woodsman and his wife as he approached the house etc. To provide the students with background inspiration for the presentation we offered among other sources the beautifully illustrated book Images of Infinity, Hemmings and Tahta (1992).

The evaluation of the project which is based upon video recordings of the student's presentations as well as the accompanying discussions following the presentations and interviews with all the students groups and all the teachers involved is not yet completed. The interviews however support the immediate impressions of the teachers involved: The students found the project very engaging and challenging as well as somewhat stressing due to the relatively short period during which they had to not only absorb new ways of looking at challenging literary texts and thought provoking mathematical problems but also making a creative presentation. These analyses also clearly point to ways of improving the goal of the project: Making students more aware of the cognitive mechanisms they use for learning mathematics, in particular the role of multiple representations in problem solving - and similarly goals for literature. Finally the preliminary analysis clearly points to the indispensability of having access to a versatile user friendly tool such as TI-*N*spire™ CAS for easy construction and interactive manipulation of multiple mathematical representations. It is very difficult to imagine how such a project could be conducted with truly interactive participation of the students without the aid of such a visual tool.

FUTURE RESEARCH DIRECTIONS

This year we have obtained funding from the Danish Association of Mathematics Teachers (upper secondary school) to revise our materials and organize a small conference for 5 upper secondary schools throughout Denmark which will test the cross curricular learning method and the teaching materials. It is our hope that this will consolidate the project and make it possible to publish the teaching materials including mathematical activities and suitable literary texts to be used in this kind of teaching. We also hope that this will provide us with a more systematic documentation of the student's progress in problem solving using suitable pretests and posttests as part of their homework, something that we only succeeded having for the literature part of the project this time.

REFERENCES

Austin, A. K. (1971). Problems and solutions. *Mathematics Magazine*, *44*(1), 16–19.

Borges, J. (1975). *El libro de arena*. Madrid, Spain: Alianza Editorial S.A..

Clarke, M. (2001). *The book of sand - A hypertext/puzzle*. Retrieved July 8, 2011, from http://artificeeternity.com/bookofsand/

Hadley, J., & Singmaster, D. (1992). Problems to sharpen the young. *The Mathematical Gazette*, *76*(475), 102–126. doi:10.2307/3620384.

Hemmings, R., & Tahta, D. (1992). *Images of infinity*. London: Tarquin Publications.

Lakoff, G., & Núñez, R. (2000). *Where mathematics comes from: How the embodied mind brings mathematics into being*. New York: Basic Books.

Thomson, J. F. (1954). Tasks and super-tasks. *Analysis*, *15*(1), 1–13. doi:10.1093/analys/15.1.1.

TI-. *N*spire™ CAS. (2007). *Dynamical mathematical software from Texas Instruments*. Retrieved from www.education.ti.com

Turner, M. (2010). *The riddle of the Buddhist monk*. Retrieved July 8, 2011, from http://markturner.org/blending.html

Turner, M., & Fauconnier, G. (2002). *The way we think: Conceptual blending and the mind's hidden complexities*. New York: Basic Books.

ENDNOTES

1 XXVI. PROPOSITIO DE CVRSV CANIS AC FVGA LEPORIS. Est campus, qui habet in longitudine pedes CL. In uno capite stabat canis, et in alio stabat lepus. Promouit namque canis ille post illum, scilicet leporem currere. Ast ubi ille canis faciebat in uno saltu pedes VIIII, lepus transmittebat VII. Dicat, qui uelit, quot pedes, quotque saltus canis persequendo, et lepus fugiendo, quoadusque comprehensus est, fecerunt (Bed., confecerint).

Chapter 12
String Art and Linear Iterative Systems

Samer Habre
Lebanese American University, Lebanon

ABSTRACT

String art dates back to the 19th century when it was initially invented to ease the delivery of some mathematical ideas. Since then, the art has evolved and so has its use in mathematics. In this chapter, the authors see how some of the phase portraits for 2 x 2 linear homogeneous iterative systems exhibit some artistic behavior that resembles this form of art. The investigation gives a sufficient condition for the solutions of such systems to form closed cycles. However, in other situations the cycles formed are infinite, producing some fascinating examples of string art.

INTRODUCTION

String art has its origins in the "curve stitch" activities invented by Mary Everest Boole (1832–1916) at the end of the 19th Century. String art or "mathematical embroidery" is a family of lines created by attaching strings to artfully located sets of pins forming abstract geometric patterns (Pedoe, 1976). Even though straight lines are formed by the string, yet the eyes of the observer pick a phantom curve or an "envelope". Mathematically, the envelope of a one parameter family of lines is the curve which is mutually tangent to that family (Struik, 1950). In simpler terms, strings can be woven so as to envelope various geometric figures such as a polygon (Figure 1a), the shape of a hyperbola (Figure 1b), or simply some beautiful design (Figure 1c).

DOI: 10.4018/978-1-4666-4050-4.ch012

Figure 1. (a) String art that envelops a polygon; (b) string art that envelops a hyperbola; (c) a beautiful string art design

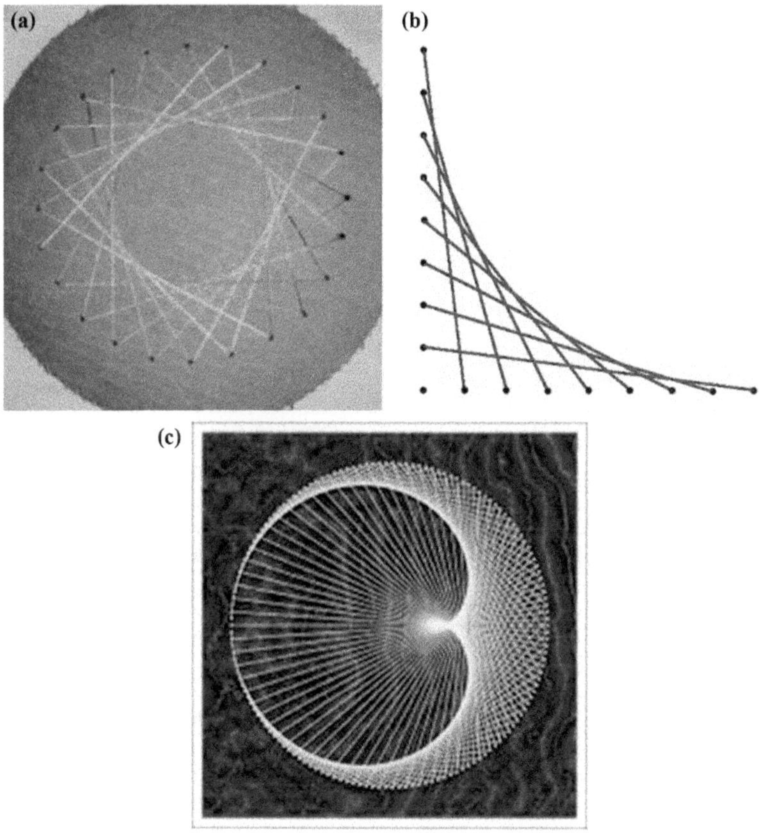

String art was initiated by Mary Boole to make mathematical ideas more accessible to children. Boole, the wife of mathematician George Boole, was a self-taught mathematician; she wrote many didactic works on mathematics, such as The Preparation of the Child for Science (Boole, 1904), and Philosophy and Fun of Algebra (Boole, 1909). The latter book encouraged children to explore mathematics through playful activities such as "curve stitching".

In the 20th century, this art was popularized as a decorative craft. There are however many instances where curve stitching is used for elementary and middle school mathematics. In more elaborate instances, string art has been used to introduce more advanced mathematical ideas such as Bézier curves. In this chapter, I present to the reader yet another mathematical instance that is associated with this form of art. The solutions of linear iterative systems as introduced in Chapter 4 of this book and when sketched using the dynamical software developed by Hu Hohn can exhibit some fascinating figures that may be classified as a form of string art.

LINEAR SYSTEMS OF ITERATIVE EQUATIONS AND STRING ART

Linear iterative systems were introduced and discussed in Chapter 4. These take the form:

$$x_{n+1} = ax_n + by_n$$
$$y_{n+1} = cx_n + dy_n$$

To understand such systems is to understand the fate of the iteration when starting with an arbitrary point in the plane (initial conditions). In all cases, the point (0,0) is an equilibrium solution in the sense that the iterated values will remain unchanged. A complete study of such systems was done by the author and it appeared in a paper entitled "A Complete Classification of 2×2 Linear Iterative Systems" (Habre & McDill, 2008). It was found that the shape and long term behavior of the solutions to the iterative systems depend on the number of eigenvalues of the system matrix $A = \begin{bmatrix} a & b \\ c & d \end{bmatrix}$.

As mentioned in Chapter 4, Hu Hohn from the Massachusetts College of Arts developed the dynamical tool *Linear Iterated Systems* to visualize the fate of any iteration. We give an example.

Example 1

Consider the system:

$$\begin{cases} x_{n+1} = -1.05x_n \\ y_{n+1} = 1.1y_n \end{cases}$$

Its general solution takes the form:

$$\vec{Y}_n = c_1\left(-1.05\right)^n \begin{bmatrix} 1 \\ 0 \end{bmatrix} + c_2\left(1.1\right)^n \begin{bmatrix} 0 \\ 1 \end{bmatrix}$$

Figure 2 is a snapshot of the phase plane of this system as well as one particular solution (initial conditions are $x_0 = -0.2; y_0 = 0.05$). As this figure shows, the iteration flips between the first and second quadrant of the plane and moves away from the origin. In this case, the equilibrium solution (0,0) is classified as a flip source (other classifications of the equilibrium solutions are found in the paper by Habre and McDill, 2008).

Figure 2 looks like strings woven in a special pattern to form a special geometric pattern. This pattern may have very well been the result of string art. In this case, the art work that results gives the visual impression that the pins to which the threads are attached outline a parabolic shape. There is one main difference though between string art and art that results from linear systems: Whereas in the former case, pins are initially fixed and then connected by a thread, in the latter case only one pin is fixed (the initial condition), and the rest are identified one at a time as a result of the iterations. The threads in the figure of the system are there only to identify the sequence of iterated points.

Figure 2. Origin is a flip source

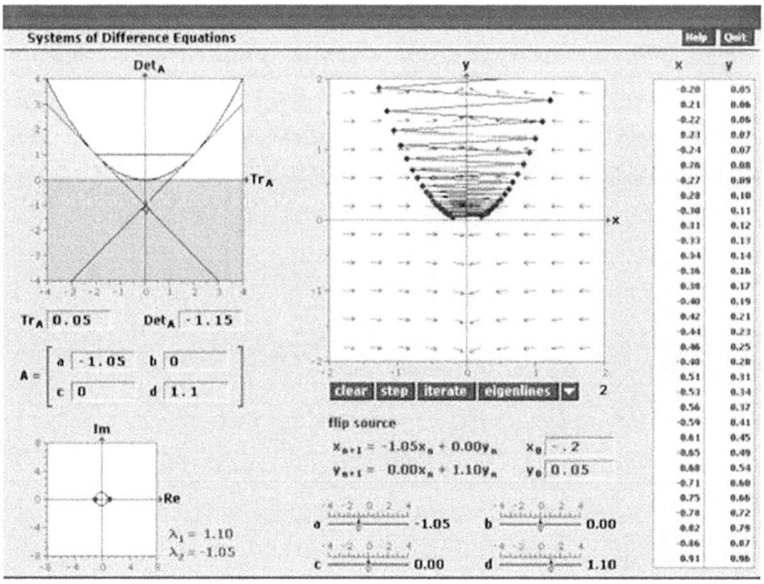

Further investigations of the linear iterative systems show that some fascinating designs can be the result of systems where the eigenvalues of the system matrix A are complex. One example was given in Chapter 4 in which case the origin was classified as a spiral source (Chapter 4, Figure 9a). Other cases are discussed next.

Example 2

Consider any linear iterative system satisfying the following conditions: The determinant of the system matrix A is 1 (ad-bc=1) and its trace ($a + d$) is bounded between -2 and 2 ($-2<a+d<2$). A random exploration of the dynamical tool developed by Hohn produced 3 special cases of cyclic iterations: Three, four, and six-point cycles around the origin (See Figures 3a, 3b, and 3c, respectively).

This led to investigate if there are other possible cycles. In fact, one can easily show that in case the system matrix is a rotation matrix of the form

Figure 3. (a) A 3-point cycle around the origin; (b) a 4-point cycle around the origin; (c) a 6-point cycle around the origin

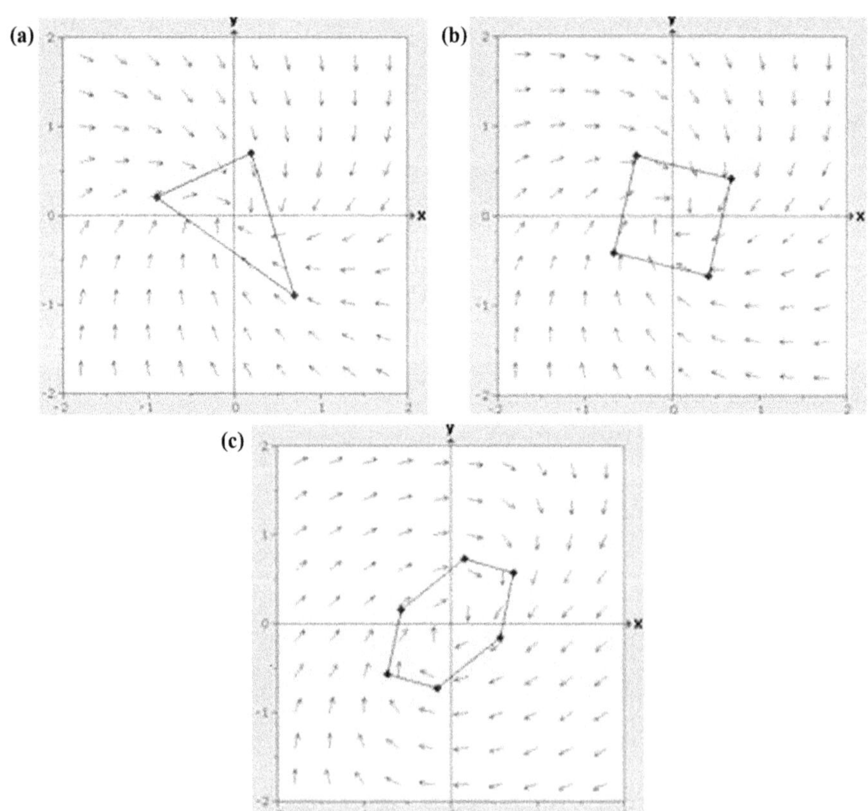

$$A = \begin{pmatrix} \cos\left(\frac{2\pi}{k}\right) & -\sin\left(\frac{2\pi}{k}\right) \\ \sin\left(\frac{2\pi}{k}\right) & \cos\left(\frac{2\pi}{k}\right) \end{pmatrix}$$

where k is a positive integer, then the iterated points form a k-point cycle.

Are the previous closed cycles in any way related to string art designs?

One of the most basic shapes in string art is the one shown in Figure 1a. It is in fact an overlap of (equilateral) triangles equally spaced along a circle. For linear systems forming 3-point cycles, overlaying cycles that are the result of various initial conditions produces a shape that resembles such a figure (See Figure 4).

A natural question now arises: Is it possible to have a linear iterative system for which the solution to a *single* initial condition resembles the string art of Figure 1a?

Additional visual explorations of the cases where the determinant of the matrix A is 1 and its trace is bounded between -2 and 2 show such systems do exist, but of course they are not the result of rotational matrices as previously defined.

Figure 4. Overlaying 3-point cycles

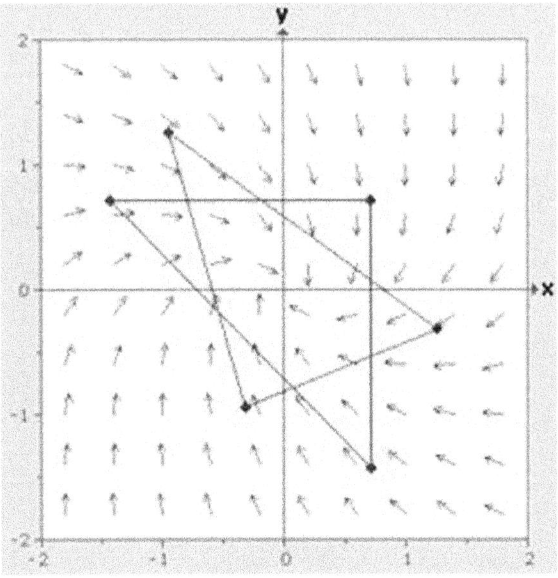

Example 3

Consider the case where the system matrix is $A = \begin{bmatrix} 0 & 1 \\ -1 & -0.28 \end{bmatrix}$ and initial conditions $x_0 = 1; y_0 = 1$. Figure 5 illustrates "many" iterations of the system. To understand further how this "art" is produced, I have shown in Figure 6a 10 iterations of the linear system, while in Figure 6b I have shown 11 and in Figure 6c 33 iterations. Calculations show that after eleven iterations, $x_{11} = 0.97, y_{11} = 1.03$ which are very close to the initial values chosen. The same is true for $x_{22}, x_{33}, \ldots, y_{22,} y_{33} \ldots$.

FURTHER EXAMPLES OF STRING ART DESIGNS

In this section, I present some further cases of infinite cycles that produce some fascinating string art designs (See Figures 7a to 7h).

CONCLUSION

In conclusion, the mathematics used in investigating linear iterative systems may be relatively simple but the outcome is a collection of beautiful art designs. Of course the software used and in particular its dynamical nature was a key component for

Figure 5. Showing many iterations; of the system with matrix $A = \begin{pmatrix} 0 & 1 \\ -1 & -0.28 \end{pmatrix}$

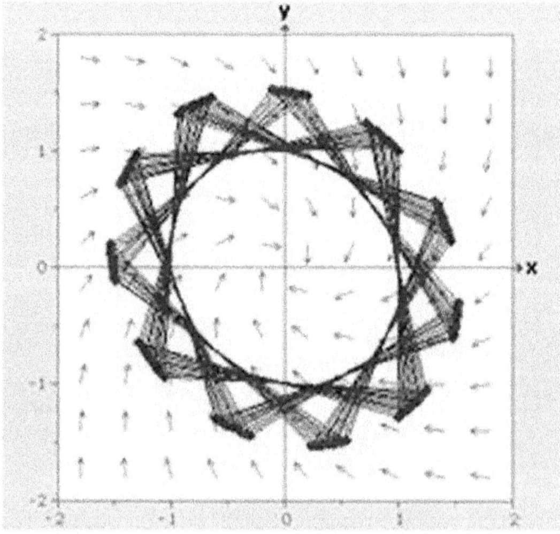

Figure 6. (a) Ten iterations around the origin; system matrix $A = \begin{pmatrix} 0 & 1 \\ -1 & -0.28 \end{pmatrix}$; *(b) eleven iterations around the origin; system matrix* $A = \begin{pmatrix} 0 & 1 \\ -1 & -0.28 \end{pmatrix}$; *(c) thirty-three iterations around the origin; system matrix* $A = \begin{pmatrix} 0 & 1 \\ -1 & -0.28 \end{pmatrix}$

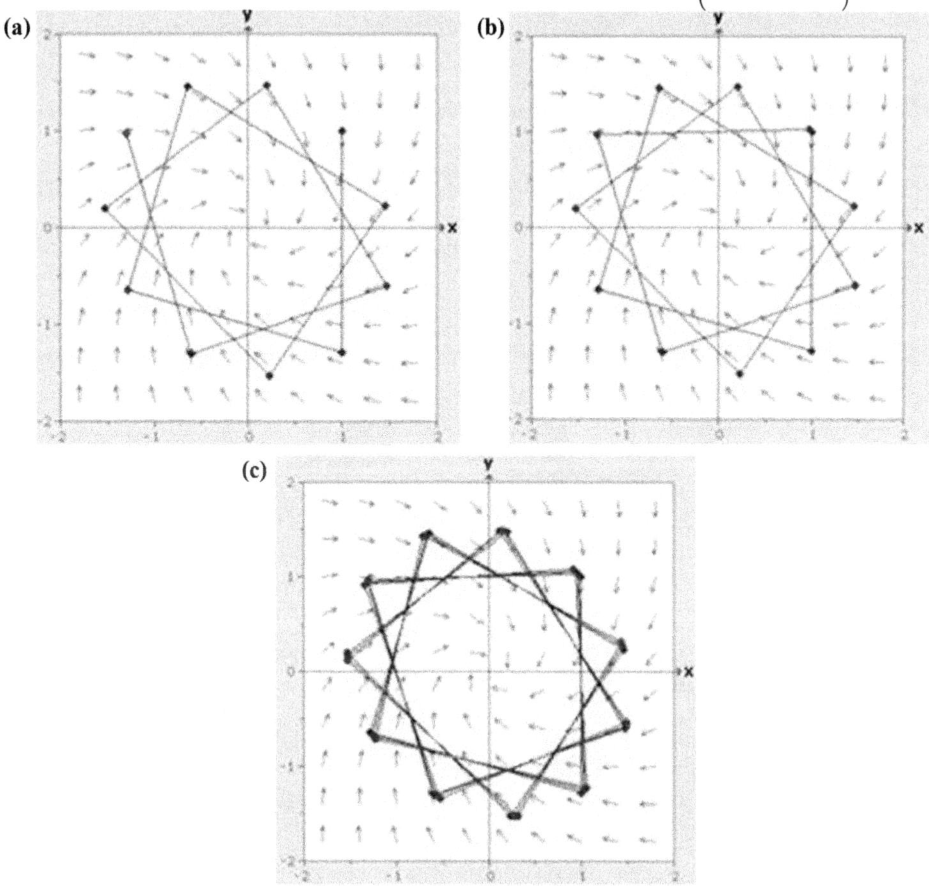

this investigation. The combination of simple mathematics, the dynamical nature of the computer program, and the beauty of string art, should encourage educators to introduce the topic of linear iterative systems to an undergraduate audience. This will not only raise awareness to a topic rarely discussed in undergraduate mathematics, but also raise eyebrows because of the artistic nature of its outcomes.

Figure 7. (a) determinant = 1; trace= -1.5; (b) determinant=1; trace = -0.9; (c) determinant=1; trace = -0.8; (d) determinant=1; trace = -0.7; (e) determinant=1; trace = 0.3; (f) determinant=1; trace = 0.6; (g) determinant=1; trace = 0.99; (h) determinant=1; trace = 1.7

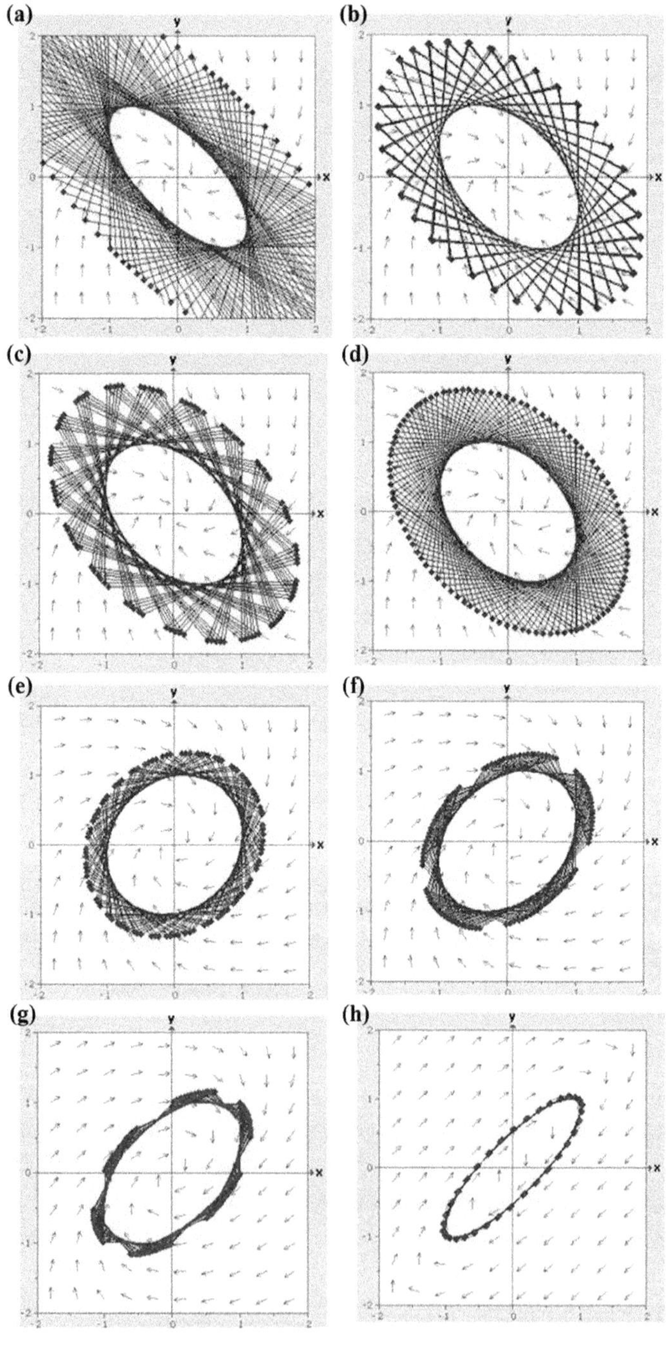

REFERENCES

Boole, M. E. (1904). *The preparation of the child for science*. Clarendon Press.

Boole, M. E. (1909). *Philosophy and fun of algebra*. C. W. Daniel.

Habre, S., & McDill, J. (2008). A complete classification of 2x2 linear iterative systems. *The Electronic Journal of Mathematics and Technology, 2*(1).

Hohn, H. (2005). *Iterated linear systems,* unpublished.

Pedoe, D. (1976). *Geometry and the visual arts*. New York: Dover.

Struik, D. J. (1950). *Lectures on classical differential geometry*. New York: Dover.

Related References

The following resources have been curated by Dr. Samer Habre to provide further insight into the research areas of visualization and mathmatics education.

Abboud-Blanchard, M. (2005). Uses of ICT by pre-service teachers. In F. Olivero & R. Sutherland (Eds.), *Proceedings of the 7th International Conference of Technology in Mathematics Teaching*, (Vol. 2 pp.74 - 78). Technology in Mathematics Teaching.

Abdoullaev, A. (2008). The world code: mathematical ontology as the real road to reality. In Abdoullaev, A. (Ed.), *Reality, Universal Ontology and Knowledge Systems: Toward the Intelligent World* (pp. 58–75). Hershey, PA: IGI Publishing. doi:10.4018/978-1-59904-966-3.ch003.

Abdoullaev, A. (2008). What organizes the world: n-relational entities. In Abdoullaev, A. (Ed.), *Reality, Universal Ontology and Knowledge Systems: Toward the Intelligent World* (pp. 128–147). Hershey, PA: IGI Publishing. doi:10.4018/978-1-59904-966-3.ch006.

Abramovitz, B., Berezina, M., Berman, A., & Shvartsman, L. (2012). A blended learning approach in mathematics. In Juan, A., Huertas, M., Trenholm, S., & Steegmann, C. (Eds.), *Teaching Mathematics Online: Emergent Technologies and Methodologies* (pp. 22–42). Hershey, PA: Information Science Reference.

Abu-Al-Aish, A., Love, S., & Hunaiti, Z. (2012). Mathematics students' readiness for mobile learning. *International Journal of Mobile and Blended Learning*, *4*(4), 1–20. doi:10.4018/jmbl.2012100101.

Abutaleb, A., & Papaioannou, M. (2006). Malliavin calculus for the estimation of the U.S. Dollar/Euro exchange rate when the volatility is stochastic. In Kurihara, Y., Takaya, S., & Yamori, N. (Eds.), *Global Information Technology and Competitive Financial Alliances* (pp. 71–101). Hershey, PA: Idea Group Publishing. doi:10.4018/978-1-59140-881-9.ch005.

Ahmed, A., & Elfessi, A. (2008). Student decision making in technology application. In Kidd, T., & Song, H. (Eds.), *Handbook of Research on Instructional Systems and Technology* (pp. 155–167). Hershey, PA: Information Science Reference. doi:10.4018/978-1-59904-865-9.ch012.

Aiken, L. R. Jr. (1997). Mathematics as a creative language. *The Arithmetic Teacher*, *24*, 251–255.

Related References

Albano, G. (2012). Mathematics education: Teaching and learning opportunities in blended learning. In Juan, A., Huertas, M., Trenholm, S., & Steegmann, C. (Eds.), *Teaching Mathematics Online: Emergent Technologies and Methodologies* (pp. 60–89). Hershey, PA: Information Science Reference.

Albano, G., & Ferrari, P. L. (2008). Integrating technology and research in mathematics education: The case of e-learning. In García-Peñalvo, F. (Ed.), *Advances in E-Learning: Experiences and Methodologies* (pp. 132–148). Hershey, PA: Information Science Reference. doi:10.4018/978-1-59904-756-0.ch008.

Ali, D. F., Patil, A., & Nordin, M. S. (2012). Visualization skills in engineering education: Issues, developments, and enhancement. In Patil, A., Eijkman, H., & Bhattacharyya, E. (Eds.), *New Media Communication Skills for Engineers and IT Professionals: Trans-National and Trans-Cultural Demands* (pp. 175–203). Hershey, PA: Information Science Reference. doi:10.4018/978-1-4666-0243-4.ch011.

Arnberger, A., & Reichhart, T. (2010). Does computer game experience influence visual scenario assessment of urban recreational paths? A case study using 3-D computer animation. In Silva, C. (Ed.), *Handbook of Research on E-Planning: ICTs for Urban Development and Monitoring* (pp. 103–119). Hershey, PA: Information Science Reference. doi:10.4018/978-1-61520-929-3.ch006.

Asif, U., & Iqbal, J. (2011). Modeling, simulation and motion cues visualization of a six-DOF motion platform for micro-manipulations. [IJIMR]. *International Journal of Intelligent Mechatronics and Robotics*, *1*(3), 1–17. doi:10.4018/ijimr.2011070101.

Asif, U., & Iqbal, J. (2013). Modeling, simulation and motion cues visualization of a six-DOF motion platform for micro-manipulations. In Sirouspour, S. (Ed.), *Advanced Engineering and Computational Methodologies for Intelligent Mechatronics and Robotics* (pp. 137–151). Hershey, PA: Information Science Reference. doi:10.4018/978-1-4666-3634-7.ch010.

Aspinwall, L., Shaw, K., & Presmeg, N. (1997). Uncontrollable mental imagery: Graphical connections between a function and its derivatives. *Educational Studies in Mathematics*, *33*, 301–317. doi:10.1023/A:1002976729261.

Aurum, A., Handzic, M., & Gardiner, A. (2008). Supporting creativity in software development: An application in IT education. In Tomei, L. (Ed.), *Online and Distance Learning: Concepts, Methodologies, Tools, and Applications* (pp. 2215–2223). Hershey, PA: Information Science Reference.

Azzag, H., Picarougne, F., Guinot, C., & Venturini, G. (2009). VRMiner: A tool for multimedia database mining with virtual reality. In Erickson, J. (Ed.), *Database Technologies: Concepts, Methodologies, Tools, and Applications* (pp. 1151–1167). Hershey, PA: Information Science Reference. doi:10.4018/978-1-60566-058-5.ch066.

Bakar, M., & Tall, D. (1991). Students' mental prototypes for functions and graphs. In Furinghetti & Fulvia (Eds.), *Proceedings of the 15th Conference of the International Group for the Psychology of Mathematics Education*, (vol. 1, pp. 104 – 111). Psychology of Mathematics Eduction.

Banas, J. R., & Brown, C. A. (2012). Web 2.0 visualization tools to stimulate generative learning. In Polly, D., Mims, C., & Persichitte, K. (Eds.), *Developing Technology-Rich Teacher Education Programs: Key Issues* (pp. 77–90). Hershey, PA: Information Science Reference. doi:10.4018/978-1-4666-0014-0.ch006.

Bardelle, C. (2013). Improving learning strategies for mathematics through e-learning. In Pumilia-Gnarini, P., Favaron, E., Pacetti, E., Bishop, J., & Guerra, L. (Eds.), *Handbook of Research on Didactic Strategies and Technologies for Education: Incorporating Advancements* (pp. 673–680). Hershey, PA: Information Science Reference.

Batra, D., VanderMeer, D., & Dutta, K. (2013). Extending agile principles to larger, dynamic software projects: A theoretical assessment. In Siau, K. (Ed.), *Innovations in Database Design, Web Applications, and Information Systems Management* (pp. 410–429). Hershey, PA: Information Science Reference.

Bellou, J. (2009). Using dynamic visualizations to enhance learning in physical geography. In Cartelli, A., & Palma, M. (Eds.), *Encyclopedia of Information Communication Technology* (pp. 795–808). Hershey, PA: Information Science Reference.

Ben Youssef, B. (2013). Integration of a visualization solution with a 3-D simulation model for tissue growth. In Rückemann, C. (Ed.), *Integrated Information and Computing Systems for Natural, Spatial, and Social Sciences* (pp. 388–407). Hershey, PA: Information Science Reference.

Bescherer, C., Herding, D., Kortenkamp, U., Müller, W., & Zimmermann, M. (2012). E-learning tools with intelligent assessment and feedback for mathematics study. In S. Graf, F. Lin, Kinshuk, & R. McGreal (Eds.), Intelligent and Adaptive Learning Systems: Technology Enhanced Support for Learners and Teachers (pp. 151-163). Hershey, PA: Information Science Reference. doi: doi:10.4018/978-1-60960-842-2.ch010.

Bessis, N., Brown, A., & Asimakopoulou, E. (2012). A mathematical analysis of a disaster management data-grid push service. In Bessis, N. (Ed.), *Technology Integration Advancements in Distributed Systems and Computing* (pp. 268–282). Hershey, PA: Information Science Reference. doi:10.4018/978-1-4666-0906-8.ch016.

Biehler, R., Fischer, P. R., Hochmuth, R., & Wassong, T. (2012). Self-regulated learning and self assessment in online mathematics bridging courses. In Juan, A., Huertas, M., Trenholm, S., & Steegmann, C. (Eds.), *Teaching Mathematics Online: Emergent Technologies and Methodologies* (pp. 216–237). Hershey, PA: Information Science Reference.

Blanchard, P. (1994). Teaching differential equations with a dynamical systems viewpoint. *The College Mathematics Journal, 25*(5), 385–395. doi:10.2307/2687503.

Borwein, J. (2005). The experimental mathematician: The pleasure of discovery and the role of proof. *International Journal of Computers for Mathematical Learning, 10*, 75–108. doi:10.1007/s10758-005-5216-x.

Related References

Brescia, W., & Cline, T. (2008). Online calculator training in mathematics and technology. In Clarke, S. (Ed.), *End-User Computing: Concepts, Methodologies, Tools, and Applications* (pp. 1–1). Hershey, PA: Information Science Reference. doi:10.4018/978-1-59904-945-8.ch094.

Brinkman, W., Rae, A., & Dwivedi, Y. K. (2007). Web-based implementation of the personalised system of instruction: A case study of teaching mathematics in an online learning environment. *International Journal of Web-Based Learning and Teaching Technologies*, *2*(1), 39–69. doi:10.4018/jwltt.2007010103.

Brinkman, W., Rae, A., & Dwivedi, Y. K. (2009). Computer-aided personalised system of instruction for teaching mathematics in an online learning environment. In Karacapilidis, N. (Ed.), *Solutions and Innovations in Web-Based Technologies for Augmented Learning: Improved Platforms, Tools, and Applications* (pp. 271–299). Hershey, PA: Information Science Reference. doi:10.4018/978-1-60566-238-1.ch017.

Brinkman, W., Rae, A., & Dwivedi, Y. K. (2010). Web-based implementation of the personalised system of instruction: A case study of teaching mathematics in an online learning environment. In Association, I. (Ed.), *Web-Based Education: Concepts, Methodologies, Tools and Applications* (pp. 1075–1101). Hershey, PA: Information Science Reference.

Bromuri, S., Urovi, V., & Stathis, K. (2012). iCampus: A connected campus in the ambient event calculus. In K. Curran (Ed.), Innovative Applications of Ambient Intelligence: Advances in Smart Systems (pp. 58-64). Hershey, PA: Information Science Reference. doi: doi:10.4018/978-1-4666-0038-6.ch005.

Brunsden, V. W. (2009). Inverting the remedial mathematics classroom with alternative assessment. In Schreiner, C. (Ed.), *Handbook of Research on Assessment Technologies, Methods, and Applications in Higher Education* (pp. 205–212). Hershey, PA: Information Science Reference. doi:10.4018/978-1-60566-667-9.ch012.

Cao, J., Zhang, F., Xu, K., & Liu, L. (2009). From enabling to ensuring grid workflows. In Wang, L., Chen, J., & Jie, W. (Eds.), *Quantitative Quality of Service for Grid Computing: Applications for Heterogeneity, Large-Scale Distribution, and Dynamic Environments* (pp. 46–73). Hershey, PA: Information Science Reference. doi:10.4018/978-1-60566-370-8.ch003.

Carvalho, P., Palhares, P., Osório, A., Gomes, A., & Mamede, E. (2012). LOGO and elementary mathematics education in Portugal. In Lennex, L., & Nettleton, K. (Eds.), *Cases on Inquiry through Instructional Technology in Math and Science* (pp. 487–522). Hershey, PA: Information Science Reference. doi:10.4018/978-1-4666-0068-3.ch018.

Castro, P., & Lopes, A. (2010). Magnet mail: A visualization system for emails. *International Journal of Creative Interfaces and Computer Graphics*, *1*(2), 29–39. doi:10.4018/jcicg.2010070103.

Castro, P., & Lopes, A. (2012). Magnet mail: A visualization system for emails. In Falchuk, B., & Marcos, A. (Eds.), *Innovative Design and Creation of Visual Interfaces: Advancements and Trends* (pp. 107–118). Hershey, PA: Information Science Reference. doi:10.4018/978-1-4666-0285-4.ch008.

Cerrito, P. (2010). Data visualization and data summary. In Cerrito, P. (Ed.), *Text Mining Techniques for Healthcare Provider Quality Determination: Methods for Rank Comparisons* (pp. 35–72). Hershey, PA: Medical Information Science Reference.

Chambel, T., & Martinho, J. (2012). Interactive visualization and exploration of video spaces through colors in motion. In Lugmayr, A., Franssila, H., Näränen, P., Sotamaa, O., Vanhala, J., & Yu, Z. (Eds.), *Media in the Ubiquitous Era: Ambient, Social and Gaming Media* (pp. 171–187). Hershey, PA: Information Science Reference.

Champney, D., & Kuo, E. (2012). An evolovong visual image of approximation with Taylor series: A case study. In *Proceedings of the 15th Annual Conference on Research in Undergraduate Mathematics Education*. Undergraduate Mathematics Education.

Chandra Pati, U. (2012). *3-D surface geometry and reconstruction: Developing concepts and applications*. Hershey, PA: IGI Global. doi:10.4018/978-1-46660-113-0.

Cherkas, B., & Welder, R. M. (2012). Interactive web-based tools for learning mathematics: Best practices. In Juan, A., Huertas, M., Trenholm, S., & Steegmann, C. (Eds.), *Teaching Mathematics Online: Emergent Technologies and Methodologies* (pp. 274–306). Hershey, PA: Information Science Reference.

Chinnappan, M. (2008). Reconceptualisation of learning objects as meta-schemas. In Lockyer, L., Bennett, S., Agostinho, S., & Harper, B. (Eds.), *Handbook of Research on Learning Design and Learning Objects: Issues, Applications, and Technologies* (pp. 792–807). Hershey, PA: Information Science Reference. doi:10.4018/978-1-59904-861-1.ch039.

Christophoridis, C., Bizani, E., & Fytianos, K. (2011). Environmental quality monitoring, using GIS as a tool of visualization, management and decision-making: Applications emerging from the EU water framework directive EU 2000/60. In Andreopoulou, Z., Manos, B., Polman, N., & Viaggi, D. (Eds.), *Agricultural and Environmental Informatics, Governance and Management: Emerging Research Applications* (pp. 397–424). Hershey, PA: Information Science Reference. doi:10.4018/978-1-60960-621-3.ch021.

Ciulla, C. (2009). Equating the two intensity-curvature terms, before and after interpolation, attempting to obtain resilient interpolation: Trivariate linear interpolation function. In Ciulla, C. (Ed.), *Improved Signal and Image Interpolation in Biomedical Applications: The Case of Magnetic Resonance Imaging (MRI)* (pp. 206–212). Hershey, PA: Medical Information Science Reference. doi:10.4018/978-1-60566-202-2.ch012.

Ciulla, C. (2009). The conception of the intensity-curvature functional. In Ciulla, C. (Ed.), *Improved Signal and Image Interpolation in Biomedical Applications: The Case of Magnetic Resonance Imaging (MRI)* (pp. 31–39). Hershey, PA: Medical Information Science Reference. doi:10.4018/978-1-60566-202-2.ch003.

Ciulla, C. (2009). The conception of the sub-pixel efficacy region. In Ciulla, C. (Ed.), *Improved Signal and Image Interpolation in Biomedical Applications: The Case of Magnetic Resonance Imaging (MRI)* (pp. 40–47). Hershey, PA: Medical Information Science Reference. doi:10.4018/978-1-60566-202-2.ch004.

Related References

Ciulla, C. (2009). The extension of the theory to the trivariate linear interpolation function. In Ciulla, C. (Ed.), *Improved Signal and Image Interpolation in Biomedical Applications: The Case of Magnetic Resonance Imaging (MRI)* (pp. 180–187). Hershey, PA: Medical Information Science Reference. doi:10.4018/978-1-60566-202-2.ch010.

Ciulla, C. (2009). The extension of theory and methodology to b-splines. In Ciulla, C. (Ed.), *Improved Signal and Image Interpolation in Biomedical Applications: The Case of Magnetic Resonance Imaging (MRI)* (pp. 223–238). Hershey, PA: Medical Information Science Reference. doi:10.4018/978-1-60566-202-2.ch014.

Ciulla, C. (2009). The theoretical approach to the improvement of the interpolation error: Bivariate linear interpolation function. In Ciulla, C. (Ed.), *Improved Signal and Image Interpolation in Biomedical Applications: The Case of Magnetic Resonance Imaging (MRI)* (pp. 58–71). Hershey, PA: Medical Information Science Reference. doi:10.4018/978-1-60566-202-2.ch007.

Ciulla, C. (2009). The unifying theory embraces lagrange and sinc interpolation functions. In Ciulla, C. (Ed.), *Improved Signal and Image Interpolation in Biomedical Applications: The Case of Magnetic Resonance Imaging (MRI)* (pp. 353–370). Hershey, PA: Medical Information Science Reference. doi:10.4018/978-1-60566-202-2.ch018.

Climescu-Haulica, A., & Quirk, M. (2010). Nonlinear stochastic differential equations method for reverse engineering of gene regulatory network. In Das, S., Caragea, D., Welch, S., & Hsu, W. (Eds.), *Handbook of Research on Computational Methodologies in Gene Regulatory Networks* (pp. 219–243). Hershey, PA: Medical Information Science Reference.

Combs, R. M., & Mazur, J. (2013). 3D modeling in a high school computer visualization class: Enacting a productive, distributed social learning environment. In Nettleton, K., & Lennex, L. (Eds.), *Cases on 3D Technology Application and Integration in Education* (pp. 359–387). Hershey, PA: Information Science Reference. doi:10.4018/978-1-4666-2815-1.ch015.

Constant, J. (2012). Digital approaches to visualization of geometric problems in wooden sangaku tablets. In Ursyn, A. (Ed.), *Biologically-Inspired Computing for the Arts: Scientific Data through Graphics* (pp. 240–253). Hershey, PA: Information Science Reference. doi:10.4018/978-1-4666-0942-6.ch013.

Coronato, A., Gallo, L., & De Pietro, G. (2011). Medical data visualization via a pervasive multi-agent platform. In Guah, M. (Ed.), *Healthcare Delivery Reform and New Technologies: Organizational Initiatives* (pp. 331–344). Hershey, PA: Information Science Reference.

Costabile, F. A., & Serpe, A. (2013). Computer-based mathematics instructions with MATCOS: A pedagogical experiment. In Pumilia-Gnarini, P., Favaron, E., Pacetti, E., Bishop, J., & Guerra, L. (Eds.), *Handbook of Research on Didactic Strategies and Technologies for Education: Incorporating Advancements* (pp. 724–738). Hershey, PA: Information Science Reference.

Crespo, S., Melfi, V., Fisch, S. M., Lesh, R. A., & Motoki, E. (2011). Television, games, and mathematics: Effects of children's interactions with multiple media. *International Journal of Gaming and Computer-Mediated Simulations*, *3*(3), 1–18. doi:10.4018/jgcms.2011070101.

Crespo, S., Melfi, V., Fisch, S. M., Lesh, R. A., & Motoki, E. (2013). Television, games, and mathematics: Effects of children's interactions with multiple media. In Ferdig, R. (Ed.), *Design, Utilization, and Analysis of Simulations and Game-Based Educational Worlds* (pp. 124–141). Hershey, PA: Information Science Reference. doi:10.4018/978-1-4666-4018-4.ch008.

Crisan, C. (2005). Mathematics teachers' own learning experiences with ICT: Shaping teachers' personal pedagogical expertise with the technology. In F. Olivero & R. Sutherland (Eds.), *Proceedings of the 7th International Conference of Technology in Mathematics Teaching*, (Vol. 2, pp. 88 – 95). Technology in Mathematics Teaching.

Cuzzocrea, A., & Mansmann, S. (2009). OLAP visualization: Models, issues, and techniques. In Wang, J. (Ed.), *Encyclopedia of Data Warehousing and Mining* (2nd ed., pp. 1439–1446). Hershey, PA: Information Science Reference.

d'Auriol, B. J., Lee, S., & Lee, Y. (2012). Visualizations of wireless sensor network data. In I. Management Association (Ed.), Wireless Technologies: Concepts, Methodologies, Tools and Applications (pp. 505-523). Hershey, PA: Information Science Reference. doi: doi:10.4018/978-1-61350-101-6.ch302.

Daher, W. (2010). Mathematics learning community flourishes in the cellular phone environment. *International Journal of Mobile and Blended Learning*, *2*(2), 1–17. doi:10.4018/jmbl.2010040101.

Daher, W. (2010). Wiki interaction tracks in geometry learning. *International Journal of E-Adoption*, *2*(4), 15–34. doi:10.4018/jea.2010100102.

Daher, W. (2011). Working in a wiki environment: Preservice teachers' experiences and perceptions: The case of geometry. In Huffman, S., Albritton, S., Wilmes, B., & Rickman, W. (Eds.), *Cases on Building Quality Distance Delivery Programs: Strategies and Experiences* (pp. 217–231). Hershey, PA: Information Science Reference.

Daher, W. (2012). Mathematics learning community flourishes in the cellular phone environment. In Parsons, D. (Ed.), *Refining Current Practices in Mobile and Blended Learning: New Applications* (pp. 137–155). Hershey, PA: Information Science Reference. doi:10.4018/978-1-4666-0053-9.ch010.

Daher, W. (2012). Wiki interaction tracks in geometry learning. In Sharma, S. (Ed.), *E-Adoption and Technologies for Empowering Developing Countries: Global Advances* (pp. 185–204). Hershey, PA: Information Science Reference. doi:10.4018/978-1-4666-0041-6.ch014.

Daher, W., & Baya'a, N. (2011). Building multimedia and web resources for teaching mathematical concepts through their historical development. In Styliaras, G., Koukopoulos, D., & Lazarinis, F. (Eds.), *Handbook of Research on Technologies and Cultural Heritage: Applications and Environments* (pp. 370–391). Hershey, PA: Information Science Reference.

dAuriol, B. J., Lee, S., & Lee, Y. (2010). Visualizations of wireless sensor network data. In H. Jin, & W. Jiang (Eds.), *Handbook of Research on Developments and Trends in Wireless Sensor Networks: From Principle to Practice* (pp. 353-370). Hershey, PA: Information Science Reference. doi:10.4018/978-1-61520-701-5.ch016

Related References

de Freitas, E. (2012). The emotional labor of imagining otherwise: undoing the mastery model of mathematics teacher identity. In Faulkner, J. (Ed.), *Disrupting Pedagogies in the Knowledge Society: Countering Conservative Norms with Creative Approaches* (pp. 174–185). Hershey, PA: Information Science Reference.

Delafontaine, M., Chavoshi, S. H., Cohn, A. G., & Van de Weghe, N. (2012). A qualitative trajectory calculus to reason about moving point objects. In Hazarika, S. (Ed.), *Qualitative Spatio-Temporal Representation and Reasoning: Trends and Future Directions* (pp. 147–167). Hershey, PA: Information Science Reference. doi:10.4018/978-1-61692-868-1.ch004.

Diehl, A., & Delrieux, C. (2012). Applications of serious games in geovisualization. In Cruz-Cunha, M. (Ed.), *Handbook of Research on Serious Games as Educational, Business and Research Tools* (pp. 25–46). Hershey, PA: Information Science Reference. doi:10.4018/978-1-4666-0149-9.ch002.

Dieterle, E., Dede, C., & Schrier, K. (2009). "Neomillennial" learning styles propagated by wireless handheld devices. In Kock, N. (Ed.), *E-Collaboration: Concepts, Methodologies, Tools, and Applications* (pp. 626–650). Hershey, PA: Information Science Reference. doi:10.4018/978-1-60566-652-5.ch049.

Divjak, B. (2012). Implementation of learning outcomes in mathematics for non-mathematics major by using e-learning. In Juan, A., Huertas, M., Trenholm, S., & Steegmann, C. (Eds.), *Teaching Mathematics Online: Emergent Technologies and Methodologies* (pp. 119–140). Hershey, PA: Information Science Reference.

Doerr, H. M., & Zangor, R. (2000). Creating meaning for and with the graphing calculator. *Educational Studies in Mathematics*, *18*(2), 13–20.

Doloc-Mihu, A. (2011). Interactive visualization tool for analysis of large image databases. In Zhang, Q., Segall, R., & Cao, M. (Eds.), *Visual Analytics and Interactive Technologies: Data, Text and Web Mining Applications* (pp. 266–281). Hershey, PA: Information Science Reference.

Downing, K. F., & Holtz, J. K. (2008). Online mathematics and physical science (mathematics, astronomy, chemistry and physics). In Downing, K., & Holtz, J. (Eds.), *Online Science Learning: Best Practices and Technologies* (pp. 216–241). Hershey, PA: Information Science Publishing. doi:10.4018/978-1-59904-986-1.ch010.

Dreyfus, T., & Eisenberg, T. (1990). On difficulties with diagrams: Theoretical issues. In *Proceedings of the 14th International Conference for the Psychology of Mathematics Education*, (vol. 1, pp. 27-34). Psychology of Mathematics Education.

Drivet, A. (2013). The toolbox: Objects and tools for doing mathematics. In Pumilia-Gnarini, P., Favaron, E., Pacetti, E., Bishop, J., & Guerra, L. (Eds.), *Handbook of Research on Didactic Strategies and Technologies for Education: Incorporating Advancements* (pp. 783–791). Hershey, PA: Information Science Reference.

Dubinsky, E. (1994). A theory and practice of learning college mathematics. In Shoenfeld, A. (Ed.), *Mathematical thinking and problem solving* (pp. 221–243). Hillsdale, NJ: Lawrence Erlbaum.

Dugdale, S. (1993). Functions and graphs – Perspectives on student thinking. In Carpenter, T. P. (Ed.), *Integrating research on the graphical representation of functions* (pp. 101–130). Hillsdale, NJ: Erlbaum.

Eilam, B., & Gurtler, O. (2010). Interactive hypermedia-based learning environment: Models of making sense of dynamic visualization. In Mukerji, S., & Tripathi, P. (Eds.), *Cases on Technological Adaptability and Transnational Learning: Issues and Challenges* (pp. 244–263). Hershey, PA: Information Science Reference. doi:10.4018/978-1-61520-779-4.ch013.

Eisenberg, T., & Dreyfus, T. (1986). On visual versus analytical thinking in mathematics. I C. Hoyles, R. Noss, & R. Sutherland (Eds.), *Proceedings of the Tenth International Conference on the Psychology of Mathematics Education*, (vol. 1, pp. 153-158). Psychology of Mathematics Education.

Eisenberg, T., & Dreyfus, T. (1991). On the reluctance to visualize in mathematics. In Zimmermann, W., & Cunningham, S. (Eds.), *Visualization in Teaching and Learning Mathematics* (pp. 25–37). Washington, DC: Mathematical Association of America.

Eley, P. M. (2013). Technology in mathematics education: A catalyst for diversity leadership. In Lewis, J., Green, A., & Surry, D. (Eds.), *Technology as a Tool for Diversity Leadership: Implementation and Future Implications* (pp. 219–229). Hershey, PA: Information Science Reference.

Elias, R. (2012). Projective geometry for 3D modeling of objects. In U. Chandra Pati (Ed.), 3-D Surface Geometry and Reconstruction: Developing Concepts and Applications (pp. 28-48). Hershey, PA: Information Science Reference. doi: doi:10.4018/978-1-4666-0113-0.ch002.

Engel, J., & Döllner, J. (2012). Immersive visualization of virtual 3D city models and its applications in e-planning. *International Journal of E-Planning Research, 1*(4), 17–34. doi:10.4018/ijepr.2012100102.

Eppler, M. J., & Burkhard, R. A. (2011). Knowledge visualization. In Schwartz, D., & Te'eni, D. (Eds.), *Encyclopedia of Knowledge Management* (2nd ed., pp. 987–999). Hershey, PA: Information Science Reference.

Er, S. N. (2013). Homeschooling meets virtual schools: Students and parent perceptions of online mathematics classes. In Yang, H., & Wang, S. (Eds.), *Cases on Formal and Informal E-Learning Environments: Opportunities and Practices* (pp. 166–184). Hershey, PA: Information Science Reference.

Escudero-Viladoms, N., & Sancho-Vinuesa, T. (2010). Analysis of interactions through a web annotation tool in a pre-university mathematics online course. In Lambropoulos, N., & Romero, M. (Eds.), *Educational Social Software for Context-Aware Learning: Collaborative Methods and Human Interaction* (pp. 79–97). Hershey, PA: Information Science Reference.

Related References

Evans, M. (2011). I'd rather be playing calculus: Adapting entertainment game structures to educational games. In Felicia, P. (Ed.), *Handbook of Research on Improving Learning and Motivation through Educational Games: Multidisciplinary Approaches* (pp. 153–175). Hershey, PA: Information Science Reference. doi:10.4018/978-1-60960-495-0.ch008.

Fan, F. (2009). A data visualization and interpretation system for sensor networks. In Erickson, J. (Ed.), *Database Technologies: Concepts, Methodologies, Tools, and Applications* (pp. 1568–1580). Hershey, PA: Information Science Reference. doi:10.4018/978-1-60566-058-5.ch095.

Farooq, H., Zakaria, N., & Siddique, M. T. (2012). An interactive visualization of genetic algorithm on 2-D graph. *International Journal of Software Science and Computational Intelligence*, *4*(1), 34–54. doi:10.4018/jssci.2012010102.

Fey, J. T. (1989). Technology and mathematics education: A survey of recent developments and important problems. *Educational Studies in Mathematics*, *20*(3), 237–272. doi:10.1007/BF00310873.

Fornara, N., & Colombetti, M. (2009). Specifying artificial institutions in the event calculus. In Dignum, V. (Ed.), *Handbook of Research on Multi-Agent Systems: Semantics and Dynamics of Organizational Models* (pp. 335–366). Hershey, PA: Information Science Reference. doi:10.4018/978-1-60566-256-5.ch014.

Foutsitzis, C. G., & Demetriadis, S. (2013). Scripted collaboration to leverage the impact of algorithm visualization tools in online learning: results from two small scale studies. *International Journal of e-Collaboration*, *9*(1), 42–56. doi:10.4018/jec.2013010104.

Fragoudakis, C., & Karampatsis, M. (2012). Visualizing content for computational geometry courses. In Lennex, L., & Nettleton, K. (Eds.), *Cases on Inquiry through Instructional Technology in Math and Science* (pp. 581–604). Hershey, PA: Information Science Reference. doi:10.4018/978-1-4666-0068-3.ch021.

Francischetti-Corrêa, M. (2013). Molecular visualization with supports of interaction, immersion, and collaboration among geographically separated research groups. In Martinho, R., Rijo, R., Cruz-Cunha, M., & Varajão, J. (Eds.), *Information Systems and Technologies for Enhancing Health and Social Care* (pp. 252–269). Hershey, PA: Medical Information Science Reference. doi:10.4018/978-1-4666-3667-5.ch017.

Fregola, C. (2011). Epistemological framework and mathematical learning. In Piu, A., & Fregola, C. (Eds.), *Simulation and Gaming for Mathematical Education: Epistemology and Teaching Strategies* (pp. 1–14). Hershey, PA: Information Science Reference.

Freitag, M. (1997). Reading and writing in the mathematics classroom. *The Mathematics Educator*, *8*, 16–21.

Gagliardi, M. (2013). Relevance of mesh dimension optimization, geometry simplification and discretization accuracy in the study of mechanical behaviour of bare metal stents. In Gangopadhyay, A. (Ed.), *Methods, Models, and Computation for Medical Informatics* (pp. 1–15). Hershey, PA: Medical Information Science Reference.

Garegae, K. G., & Moalosi, S. S. (2010). Botswana ICT policy and curriculum concerns: Does school connectivity guarantee technology integration into mathematics classrooms? In Adomi, E. (Ed.), *Handbook of Research on Information Communication Technology Policy: Trends, Issues and Advancements* (pp. 15–32). Hershey, PA: Information Science Reference. doi:10.4018/978-1-61520-847-0.ch002.

Gasparis, E. (2007). LePUS: A formal language for modeling design patterns. In Taibi, T. (Ed.), *Design Pattern Formalization Techniques* (pp. 357–372). Hershey, PA: Idea Group Publishing. doi:10.4018/978-1-59904-219-0.ch016.

Gesquière, G., & Manin, A. (2012). 3D visualization of urban data based on CityGML with WebGL. *International Journal of 3-D Information Modeling, 1*(3), 1-15. doi:10.4018/ij3dim.2012070101

Ghani, D. A. (2011). Visualization elements of shadow play technique movement and study of computer graphic imagery (CGI) in Wayang Kulit Kelantan. *International Journal of Art, Culture and Design Technologies, 1*(1), 50–57. doi:10.4018/IJACDT.2011010105.

Gluz, J. C., Cabral, T., Baggio, P., Livi, P., & Mallmann, R. (2008). Pedagogical negotiation and solidarity assimilation groups in action: The combined efforts of E-M@T and Leibniz to aid calculus students. In Viccari, R., Jaques, P., & Verdin, R. (Eds.), *Agent-Based Tutoring Systems by Cognitive and Affective Modeling* (pp. 260–291). Hershey, PA: Information Science Reference. doi:10.4018/978-1-59904-768-3.ch012.

Goldenberg, E. P. (1987). Believing is seeing: How preconceptions influence the perceptions of graphs. In J. Bergeron, N. Herscovits, & C. Kieran (Eds.), *Proceedings of the Eleventh Conference of the International Group for the Psychology of Mathematics Education*, (vol. 1, pp. 197-203). Psychology of Mathematics Education.

Golemati, M., Vassilakis, C., Katifori, A., Lepouras, G., & Halatsis, C. (2009). Context and adaptivity-driven visualization method selection. In Mourlas, C., & Germanakos, P. (Eds.), *Intelligent User Interfaces: Adaptation and Personalization Systems and Technologies* (pp. 188–204). Hershey, PA: Information Science Reference.

Gollapudi, S., Marshall, A., Zadik, D., & Hodgman, C. (2009). Graphical analysis and visualization tools for protein interaction networks. In Li, X., & Ng, S. (Eds.), *Biological Data Mining in Protein Interaction Networks* (pp. 286–311). Hershey, PA: Medical Information Science Reference. doi:10.4018/978-1-60566-398-2.ch016.

Greenshields, I., & El-Sayed, G. (2012). Aspects of visualization and the grid in a biomedical context. In I. Management Association (Ed.), Grid and Cloud Computing: Concepts, Methodologies, Tools and Applications (pp. 1686-1701). Hershey, PA: Information Science Reference. doi: doi:10.4018/978-1-4666-0879-5.ch710.

Greenwood, D., & Sommerville, I. (2011). Using network analysis and visualization to analyze problematic enterprise information systems. *International Journal of Knowledge and Systems Science*, *2*(4), 54–71. doi:10.4018/jkss.2011100105.

Related References

Greenwood, D., & Sommerville, I. (2013). Using network analysis and visualization to analyze problematic enterprise information systems. In Yang, G. (Ed.), *Multidisciplinary Studies in Knowledge and Systems Science* (pp. 291–310). Hershey, PA: Information Science Reference.

Habre, S. (2000). Exploring students' strategies to solve ordinary differential equations in a reformed setting. *The Journal of Mathematical Behavior, 18*(4), 455–472. doi:10.1016/S0732-3123(00)00024-9.

Habre, S. (2001). Visualization in multivariable calculus: The case of 3D-surfaces. *Focus on Learning Problems in Mathematics, 23*(1), 30–48.

Habre, S. (2003). Investigating students' approval of a geometric approach to differential equations and their solutions. *International Journal of Mathematical Education in Science and Technology, 34*(5), 651–662. doi:10.1080/0020739031000148912.

Hai-Jew, S. (2010). Information and visualization imagery. In Hai-Jew, S. (Ed.), *Digital Imagery and Informational Graphics in E-Learning: Maximizing Visual Technologies* (pp. 89–117). Hershey, PA: Information Science Reference.

Hai-Jew, S. (2011). Information and visualization imagery. In I. Management Association (Ed.), Gaming and Simulations: Concepts, Methodologies, Tools and Applications (pp. 752-778). Hershey, PA: Information Science Reference. doi: doi:10.4018/978-1-60960-195-9.ch311.

Hai-Jew, S. (2014). *Packaging digital information for enhanced learning and analysis: Data visualization, spatialization, and multidimensionality*. Hershey, PA: IGI Global.

Hall, V. G., & Martin, L. E. (1999). Making decisions about software for classroom use. *Reading Research and Instruction, 38*(3), 187–196. doi:10.1080/19388079909558288.

Hallett, D. H. (1991). Visualization and calculus reform. In Zimmermann, W., & Cunningham, S. (Eds.), *Visualization in Teaching and Learning Mathematics* (pp. 121–126). Washington, DC: Mathematical Association of America.

Hamann, J. R. (2009). Autognomic intellisite. In Khosrow-Pour, M. (Ed.), *Encyclopedia of Information Science and Technology* (2nd ed., pp. 294–299). Hershey, PA: Information Science Reference.

Hamilton-Wright, A., & Stashuk, D. W. (2009). Visualization techniques for confidence based data. In Wang, J. (Ed.), *Encyclopedia of Data Warehousing and Mining* (2nd ed., pp. 2068–2073). Hershey, PA: Information Science Reference.

Hanna, G. (2000). Proof, explanation and exploration: An overview. *Educational Studies in Mathematics, 44*, 5–23. doi:10.1023/A:1012737223465.

Harland, D. J., Pérez, Y., & Toledo, C. (2012). A call for the use of technology within mathematics and science preservice teacher methods courses. In Polly, D., Mims, C., & Persichitte, K. (Eds.), *Developing Technology-Rich Teacher Education Programs: Key Issues* (pp. 357–377). Hershey, PA: Information Science Reference. doi:10.4018/978-1-4666-0014-0.ch023.

Hartsell, T., Herron, S. S., Fang, H., & Rathod, A. (2010). Improving teachers' self-confidence in learning technology skills and math education through professional development. *International Journal of Information and Communication Technology Education*, *6*(2), 47–61. doi:10.4018/jicte.2010040105.

Hazarika, S. M. (2009). Qualitative spatial reasoning. In Khosrow-Pour, M. (Ed.), *Encyclopedia of Information Science and Technology* (2nd ed., pp. 3175–3180). Hershey, PA: Information Science Reference.

Healy, L., & Hoyles, C. (2001). Software tools for geometrical problem solving: Potentials and pitfalls. *International Journal of Computers for Mathematical Learning*, *6*, 235–256. doi:10.1023/A:1013305627916.

Heid, M. K., Sheets, C., & Matras, M. A. (1990). Computer-enhanced algebra: New roles and challenges for teachers and students. In Cooney, T. (Ed.), *Teaching and Learning Mathematics in the 1990s, NCTM 1990 Yearbook* (pp. 194–204). Reston, VA: NCTM.

Hersh, R. (1993). Proving is convincing and explaining. *Educational Studies in Mathematics*, *24*, 389–399. doi:10.1007/BF01273372.

Hoyles, C. (1998). A culture of proving in school mathematics. In Tinsley, D., & Johnson, D. (Eds.), *Information and Communication Technologies in School Mathematics* (pp. 169–181). London: Chapman and Hall. doi:10.1007/978-0-387-35287-9_21.

Huang, M., & Huang, W. (2014). *Innovative approaches of data visualization and visual analytics*. Hershey, PA: IGI Global.

Hyyppä, J., Zhu, L., Liu, Z., Kaartinen, H., & Jaakkola, A. (2012). 3D city modeling and visualization for smart phone applications. In Chen, R. (Ed.), *Ubiquitous Positioning and Mobile Location-Based Services in Smart Phones* (pp. 254–296). Hershey, PA: Information Science Reference. doi:10.4018/978-1-4666-1827-5.ch010.

Hyyppä, J., Zhu, L., Liu, Z., Kaartinen, H., & Jaakkola, A. (2013). 3D city modeling and visualization for smart phone applications. In I. Management Association (Ed.), Geographic Information Systems: Concepts, Methodologies, Tools, and Applications(pp. 1011-1052). Hershey, PA: Information Science Reference. doi: doi:10.4018/978-1-4666-2038-4.ch062.

Ikeda, H., & Ohsawa, Y. (2013). Visualization of insight process in concept creation focusing handwriting features. *International Journal of Knowledge and Systems Science*, *4*(1), 18–31. doi:10.4018/jkss.2013010102.

Incikabi, L., & Tokmak, H. S. (2013). Integrating technology into mathematics teaching: A TPACK (technological, pedagogical, content knowledge)-based course design for college students. In Keengwe, J. (Ed.), *Research Perspectives and Best Practices in Educational Technology Integration* (pp. 288–303). Hershey, PA: Information Science Reference.

Isik-Ercan, Z., Kim, B., & Nowak, J. (2012). Can 3D visualization assist in young children's understanding of sun-earth-moon system? *International Journal of Knowledge Society Research*, *3*(4), 12–21. doi:10.4018/jksr.2012100102.

Janakova, M. (2011). Visualization and simulation for the analysis of business intelligence products. *International Journal of E-Entrepreneurship and Innovation*, *2*(4), 20–31. doi:10.4018/jeei.2011100102.

Janakova, M. (2013). Visualization and simulation for the analysis of business intelligence products. In Nasir, S. (Ed.), *Modern Entrepreneurship and E-Business Innovations* (pp. 186–197). Hershey, PA: Business Science Reference.

Jarvis, D. H. (2012). Teaching mathematics teachers online: Strategies for navigating the intersection of andragogy, technology, and reform-based mathematics education. In Juan, A., Huertas, M., Trenholm, S., & Steegmann, C. (Eds.), *Teaching Mathematics Online: Emergent Technologies and Methodologies* (pp. 187–199). Hershey, PA: Information Science Reference.

Jasute, E., & Dagiene, V. (2012). Towards digital competencies in mathematics education: A model of interactive geometry. *International Journal of Digital Literacy and Digital Competence, 3*(2), 1–19. doi:10.4018/jdldc.2012040101.

Jeschke, S., & Richter, T. (2007). Mathematics in virtual knowledge spaces: User adaptation by intelligent assistants. In Kaschek, R. (Ed.), *Intelligent Assistant Systems: Concepts, Techniques and Technologies* (pp. 232–263). Hershey, PA: Idea Group Publishing. doi:10.4018/978-1-59140-878-9.ch011.

Jeschke, S., & Richter, T. (2008). Mathematics in virtual knowledge spaces: User adaptation by intelligent assistants. In Kisielnicki, J. (Ed.), *Virtual Technologies: Concepts, Methodologies, Tools, and Applications* (pp. 1582–1604). Hershey, PA: Information Science Reference. doi:10.4018/978-1-59904-955-7.ch103.

Johnston, C. J. (2010). Pre-service elementary teachers' evaluations of technology tools for mathematical learning: A reflective model. In Yamamoto, J., Kush, J., Lombard, R., & Hertzog, C. (Eds.), *Technology Implementation and Teacher Education: Reflective Models* (pp. 258–276). Hershey, PA: Information Science Reference. doi:10.4018/978-1-61520-897-5.ch015.

Johnston, C. J. (2013). Evaluation of a hybrid mathematics methods course for novice teachers. *International Journal of Online Pedagogy and Course Design, 3*(1), 33–52. doi:10.4018/ijopcd.2013010103.

Johnston, C. J., & Moyer-Packenham, P. S. (2012). A model for examining the criteria used by pre-service elementary teachers in their evaluation of technology for mathematics teaching. In Ronau, R., Rakes, C., & Niess, M. (Eds.), *Educational Technology, Teacher Knowledge, and Classroom Impact: A Research Handbook on Frameworks and Approaches* (pp. 200–227). Hershey, PA: Information Science Publishing.

Jones, B., Failla, A., & Miller, B. (2009). Tacit knowledge in rapidly evolving organisational environments. In Zaphiris, P., & Ang, C. (Eds.), *Cross-Disciplinary Advances in Human Computer Interaction: User Modeling, Social Computing, and Adaptive Interfaces* (pp. 139–158). Hershey, PA: Information Science Reference. doi:10.4018/978-1-60566-142-1.ch009.

Juan, A., Huertas, M. A., Trenholm, S., & Steegmann, C. (2012). *Teaching mathematics online: Emergent technologies and methodologies*. Hershey, PA: IGI Global.

Kacmar, C. J., Fiorito, S. S., & Carey, J. M. (2009). The influence of attitude on the acceptance and use of information systems. *Information Resources Management Journal, 22*(2), 22–49. doi:10.4018/irmj.2009040102.

Kalloo, V., & Mohan, P. (2011). An investigation into mobile learning for high school mathematics. *International Journal of Mobile and Blended Learning, 3*(3), 59–76. doi:10.4018/jmbl.2011070105.

Kalloo, V., & Mohan, P. (2013). An investigation into mobile learning for high school mathematics. In Parsons, D. (Ed.), *Innovations in Mobile Educational Technologies and Applications* (pp. 301–317). Hershey, PA: Information Science Reference.

Kaput, J. (1992). Technology and mathematics education. In Grouws, D. A. (Ed.), *Handbook of Research on Mathematics Teaching and Learning* (pp. 515–556). Reston, VA: National Council of Teachers of Mathematics.

Kashefi, H., Ismail, Z., & Yusof, Y. M. (2012). Designing a blended learning model to support mathematical thinking in multivariable calculus. In Yusof, K., Azli, N., Kosnin, A., Yusof, S., & Yusof, Y. (Eds.), *Outcome-Based Science, Technology, Engineering, and Mathematics Education: Innovative Practices* (pp. 221–239). Hershey, PA: Information Science Reference. doi:10.4018/978-1-4666-1809-1.ch011.

Keene, K. (2007). A characterization of dynamic reasoning: Reasoning with time as parameter. *The Journal of Mathematical Behavior, 26*(3), 230–246. doi:10.1016/j.jmathb.2007.09.003.

Keller, B. A., & Hircsh, C. R. (1998). Student preferences for representations of functions. *International Journal of Mathematical Education in Science and Technology, 29*, 1–17. doi:10.1080/0020739980290101.

Kelly, H. (2010). How games and simulations can help meet America's challenges in science mathematics and technology education. In Cannon-Bowers, J., & Bowers, C. (Eds.), *Serious Game Design and Development: Technologies for Training and Learning* (pp. 117–133). Hershey, PA: Information Science Reference. doi:10.4018/978-1-61520-739-8.ch007.

Ker, H. W. (2012). Engineering education and attitudes toward mathematics: A comparative study. *International Journal of Quality Assurance in Engineering and Technology Education, 2*(1), 63–76. doi:10.4018/ijqaete.2012010105.

Kersaint, G. (2010). Reflections on a course designed to encourage technology integration in secondary school mathematics. In Yamamoto, J., Kush, J., Lombard, R., & Hertzog, C. (Eds.), *Technology Implementation and Teacher Education: Reflective Models* (pp. 277–294). Hershey, PA: Information Science Reference. doi:10.4018/978-1-61520-897-5.ch016.

Kim, T., Hwang, H., & Choi, H. (2010). Cancer cell image analysis and visualization. *International Journal of E-Health and Medical Communications, 1*(1), 53–63. doi:10.4018/jehmc.2010010105.

Kim, T., Hwang, H., & Choi, H. (2012). Cancer cell image analysis and visualization. In Rodrigues, J. (Ed.), *Emerging Communication Technologies for E-Health and Medicine* (pp. 233–243). Hershey, PA: Medical Information Science Reference. doi:10.4018/978-1-4666-0909-9.ch018.

Related References

Knuth, E. J. (2000). Student understanding of the Cartesian connection: An exploratory study. *Journal for Research in Mathematics Education, 31*, 500–508. doi:10.2307/749655.

Kochar, M. (2011). User driven learning in mathematics. In Biswas, R., & Martin, C. (Eds.), *User-Driven Healthcare and Narrative Medicine: Utilizing Collaborative Social Networks and Technologies* (pp. 242–254). Hershey, PA: Medical Information Science Reference.

Kremen, P., & Kostov, B. (2012). Expressive OWL queries: Design, evaluation, visualization. *International Journal on Semantic Web and Information Systems*, *8*(4), 57–79. doi:10.4018/jswis.2012100104.

Kunkel, C. (2013). Creating and implementing a virtual math tutoring lab for undergraduate students. In Smyth, E., & Volker, J. (Eds.), *Enhancing Instruction with Visual Media: Utilizing Video and Lecture Capture* (pp. 128–137). Hershey, PA: Information Science Reference. doi:10.4018/978-1-4666-3962-1.ch009.

Laborde, C. (2000). Dynamic geometry environments as a source of rich learning contexts for the complex activity of proving. *Educational Studies in Mathematics, 44*, 151–161. doi:10.1023/A:1012793121648.

Laborde, C. (2001). Integration of technology in the design of geometry tasks with Cabri-geometry. *International Journal of Computers for Mathematical Learning*, *6*(3), 283–317. doi:10.1023/A:1013309728825.

Lammintakanen, J. (2008). Curriculum development in web-based education. In Tomei, L. (Ed.), *Online and Distance Learning: Concepts, Methodologies, Tools, and Applications* (pp. 1219–1251). Hershey, PA: Information Science Reference.

Lammintakanen, J., & Rissanen, S. (2009). Micro and macro level issues in curriculum development. In Khosrow-Pour, M. (Ed.), *Encyclopedia of Information Science and Technology* (2nd ed., pp. 2546–2550). Hershey, PA: Information Science Reference.

Lammintakanen, J., & Rissanen, S. (2009). Online learning experiences of university students. In Rogers, P., Berg, G., Boettcher, J., Howard, C., Justice, L., & Schenk, K. (Eds.), *Encyclopedia of Distance Learning* (2nd ed., pp. 1512–1518). Hershey, PA: Information Science Reference. doi:10.4018/978-1-60566-198-8.ch220.

Lansiquot, R. D. (2013). *Cases on interdisciplinary research trends in science, technology, engineering, and mathematics: Studies on urban classrooms*. Hershey, PA: IGI Global.

Leigh, J., Johnson, A., Renambot, L., Vishwanath, V., Peterka, T., & Schwarz, N. (2012). Visualization of large-scale distributed data. In Kosar, T. (Ed.), *Data Intensive Distributed Computing: Challenges and Solutions for Large-scale Information Management* (pp. 242–274). Hershey, PA: Information Science Reference. doi:10.4018/978-1-61520-971-2.ch011.

Leng, J., Rhyne, T., & Sharrock, W. (2012). Visualization: Future technology and practices for computational science and engineering. In Leng, J., & Sharrock, W. (Eds.), *Handbook of Research on Computational Science and Engineering: Theory and Practice* (pp. 381–413). Hershey, PA: Engineering Science Reference.

Leng, J., & Sharrock, W. (2010). Visualization in support of social networking on the web. In Dasgupta, S. (Ed.), *Social Computing: Concepts, Methodologies, Tools, and Applications* (pp. 2317–2336). Hershey, PA: Information Science Reference.

Li, G., Citrin, D., Miller, R. W., Camphausen, K., Mueller, B., Mychalczak, B., & Song, Y. (2011). 3D and 4D medical image registration combined with image segmentation and visualization. In I. Management Association (Ed.), Clinical Technologies: Concepts, Methodologies, Tools and Applications (pp. 885-894). Hershey, PA: Medical Information Science Reference. doi: doi:10.4018/978-1-60960-561-2.ch317.

Li, X., Xu, W., Zhu, Q., Hu, J., Hu, H., & Zhang, Y. (2012). A multi-level cache approach for realtime visualization of massive 3D GIS data. *International Journal of 3-D Information Modeling, 1*(3), 37-48. doi:10.4018/ij3dim.2012070104

Liao, H. (2013). Biomedical information processing and visualization for minimally invasive neurosurgery. In Wu, J. (Ed.), *Technological Advancements in Biomedicine for Healthcare Applications* (pp. 36–46). Hershey, PA: Medical Information Science Reference.

Lin, H. (2007). From logic specification to y-calculus: A method for designing multiagent systems. *International Journal of Intelligent Information Technologies*, *3*(3), 21–40. doi:10.4018/jiit.2007070102.

Lin, H., Kemp, J., & Gilbert, P. (2010). Computing gamma calculus on computer cluster. *International Journal of Technology Diffusion*, *1*(4), 42–52. doi:10.4018/jtd.2010100104.

Lin, H., Kemp, J., & Gilbert, P. (2012). Computing gamma calculus on computer cluster. In I. Management Association (Ed.), Computer Engineering: Concepts, Methodologies, Tools and Applications (pp. 2016-2026). Hershey, PA: Engineering Science Reference. doi: doi:10.4018/978-1-61350-456-7.ch813.

Lin, H., Kemp, J., & Gilbert, P. (2012). Computing gamma calculus on computer cluster. In Zolait, A. (Ed.), *Knowledge and Technology Adoption, Diffusion, and Transfer: International Perspectives* (pp. 275–286). Hershey, PA: Information Science Reference. doi:10.4018/978-1-4666-1752-0.ch020.

Liu, H., Gao, J., Huang, J., Beck, M., & Moore, T. (2012). On-demand visualization on scalable shared infrastructure. In Kosar, T. (Ed.), *Data Intensive Distributed Computing: Challenges and Solutions for Large-scale Information Management* (pp. 275–290). Hershey, PA: Information Science Reference. doi:10.4018/978-1-61520-971-2.ch012.

Liu, S., Eklund, T., Collan, M., & Sarlin, P. (2010). A visualization and clustering approach to analyzing the early warning signals of currency crises. In Wang, J., & Wang, S. (Eds.), *Business Intelligence in Economic Forecasting: Technologies and Techniques* (pp. 65–81). Hershey, PA: Business Science Reference. doi:10.4018/978-1-61520-629-2.ch004.

Loch, B. (2012). Screencasting for mathematics online learning: A case study of a first year operations research course at a dual delivery mode australian university. In Juan, A., Huertas, M., Trenholm, S., & Steegmann, C. (Eds.), *Teaching Mathematics Online: Emergent Technologies and Methodologies* (pp. 43–59). Hershey, PA: Information Science Reference. doi:10.4018/978-1-4666-0011-9.ch810.

Loch, B. (2012). Screencasting for mathematics online learning: A case study of a first year operations research course at a dual delivery mode Australian university. In I. Management Association (Ed.), Virtual Learning Environments: Concepts, Methodologies, Tools and Applications (pp. 1701-1716). Hershey, PA: Information Science Reference. doi: doi:10.4018/978-1-4666-0011-9.ch810.

Lokar, M., Lukšic, P., & Horvat, B. (2012). NAUK.si: Using learning blocks to prepare e-content for teaching mathematics. In Juan, A., Huertas, M., Trenholm, S., & Steegmann, C. (Eds.), *Teaching Mathematics Online: Emergent Technologies and Methodologies* (pp. 307–326). Hershey, PA: Information Science Reference.

Loukis, E., Xenakis, A., & Tserpeli, N. (2011). An investigation of the use of computer supported arguments visualization for improving public participation in legislation formation. In Cropf, R., & Krummenacher, W. (Eds.), *Information Communication Technologies and the Virtual Public Sphere: Impacts of Network Structures on Civil Society* (pp. 156–172). Hershey, PA: Information Science Reference. doi:10.4018/978-1-60960-159-1.ch008.

Lourenço, N., & Pereira, F. B. (2011). PSO-CGO: A particle swarm algorithm for cluster geometry optimization. *International Journal of Natural Computing Research*, *2*(1), 1–20. doi:10.4018/jncr.2011010101.

Lyublinskaya, I., & Tournaki, N. (2010). Preparing qualified elementary school teachers: Connecting mathematics and science by integrating data collection technology into methods courses. In Yamamoto, J., Kush, J., Lombard, R., & Hertzog, C. (Eds.), *Technology Implementation and Teacher Education: Reflective Models* (pp. 203–227). Hershey, PA: Information Science Reference. doi:10.4018/978-1-61520-897-5.ch012.

Maat, S. M., & Zakaria, E. (2012). Use of computer algebra systems in teaching and learning of ordinary differential equations among engineering technology students. In Yusof, K., Azli, N., Kosnin, A., Yusof, S., & Yusof, Y. (Eds.), *Outcome-Based Science, Technology, Engineering, and Mathematics Education: Innovative Practices* (pp. 207–220). Hershey, PA: Information Science Reference. doi:10.4018/978-1-4666-1809-1.ch010.

Maat, S. M., & Zakaria, E. (2012). Use of computer algebra systems in teaching and learning of ordinary differential equations among engineering technology students. In Yusof, K., Azli, N., Kosnin, A., Yusof, S., & Yusof, Y. (Eds.), *Outcome-Based Science, Technology, Engineering, and Mathematics Education: Innovative Practices* (pp. 207–220). Hershey, PA: Information Science Reference. doi:10.4018/978-1-4666-1809-1.ch010.

Mac an Bhaird, C., & O'Shea, A. (2012). The role of technology in mathematics support: A pilot study. In Juan, A., Huertas, M., Trenholm, S., & Steegmann, C. (Eds.), *Teaching Mathematics Online: Emergent Technologies and Methodologies* (pp. 367–383). Hershey, PA: Information Science Reference.

MacLennan, B. (2009). History of artificial intelligence before computers. In Khosrow-Pour, M. (Ed.), *Encyclopedia of Information Science and Technology* (2nd ed., pp. 1763–1768). Hershey, PA: Information Science Reference.

Maina, F., & Burrell, M. M. (2013). Attitudes towards diverse learners: A hybrid technology course for secondary school mathematics pre/in-service teachers. In Yang, H., & Wang, S. (Eds.), *Cases on Online Learning Communities and Beyond: Investigations and Applications* (pp. 113–133). Hershey, PA: Information Science Reference.

Mamykina, L., & Mynatt, E. (2009). Interpreting health and wellness information. In Symonds, J., Ayoade, J., & Parry, D. (Eds.), *Auto-Identification and Ubiquitous Computing Applications* (pp. 74–90). Hershey, PA: Information Science Reference. doi:10.4018/978-1-60566-298-5.ch005.

Marks-Tarlow, T. (2013). Fractal geometry as a bridge between realms. In Orsucci, F., & Sala, N. (Eds.), *Complexity Science, Living Systems, and Reflexing Interfaces: New Models and Perspectives* (pp. 25–43). Hershey, PA: Information Science Reference.

Masso, M. (2010). Improving prediction accuracy via subspace modeling in a statistical geometry based computational protein mutagenesis. *International Journal of Knowledge Discovery in Bioinformatics*, *1*(4), 54–68. doi:10.4018/jkdb.2010100103.

Masso, M. (2013). Improving prediction accuracy via subspace modeling in a statistical geometry based computational protein mutagenesis. In I. Management Association (Ed.), Bioinformatics: Concepts, Methodologies, Tools, and Applications (pp. 1010-1024). Hershey, PA: Medical Information Science Reference. doi: doi:10.4018/978-1-4666-3604-0.ch054.

McLaughlin, J., Fang, S., Jacobson, S. W., Hoyme, H. E., Robinson, L., & Foroud, T. (2011). Interactive feature visualization and detection for 3D face classification. *International Journal of Cognitive Informatics and Natural Intelligence*, *5*(2), 1–16. doi:10.4018/jcini.2011040101.

McLaughlin, J., Fang, S., Jacobson, S. W., Hoyme, H. E., Robinson, L., & Foroud, T. (2013). Interactive feature visualization and detection for 3D face classification. In Wang, Y. (Ed.), *Cognitive Informatics for Revealing Human Cognition: Knowledge Manipulations in Natural Intelligence* (pp. 98–110). Hershey, PA: Information Science Reference.

Meletiou-Mavrotheris, M. (2013). Integrating game-enhanced mathematics learning into the pre-service training of teachers. In de Freitas, S., Ott, M., Popescu, M., & Stanescu, I. (Eds.), *New Pedagogical Approaches in Game Enhanced Learning: Curriculum Integration* (pp. 159–179). Hershey, PA: Information Science Reference. doi:10.4018/978-1-4666-3950-8.ch009.

Related References

Meletiou-Mavrotheris, M., & Mavrotheris, E. (2010). SMASH: Blended training for parent education in mathematics and science. In Mukerji, S., & Tripathi, P. (Eds.), *Cases on Interactive Technology Environments and Transnational Collaboration: Concerns and Perspectives* (pp. 171–192). Hershey, PA: Information Science Reference. doi:10.4018/978-1-61520-909-5.ch010.

Meletiou-Mavrotheris, M., & Mavrou, K. (2013). Web-based simulations for the training of mathematics teachers. In Ritzhaupt, A., & Kumar, S. (Eds.), *Cases on Educational Technology Implementation for Facilitating Learning* (pp. 308–340). Hershey, PA: Information Science Reference. doi:10.4018/978-1-4666-3676-7.ch018.

Melis, E., Goguadze, G., Libbrecht, P., & Ullrich, C. (2010). Culturally aware mathematics education technology. In Blanchard, E., & Allard, D. (Eds.), *Handbook of Research on Culturally-Aware Information Technology: Perspectives and Models* (pp. 543–557). Hershey, PA: Information Science Reference. doi:10.4018/978-1-61520-883-8.ch025.

Miller, C., Lecheler, L., Hosack, B., Doering, A., & Hooper, S. (2012). Orchestrating data, design, and narrative: Information visualization for sense- and decision-making in online learning. *International Journal of Cyber Behavior, Psychology and Learning*, *2*(2), 1–15. doi:10.4018/ijcbpl.2012040101.

Miller, D., & Glober, D. (2010). Enhanced interactivity in secondary mathematics. In Thomas, M., & Schmid, E. (Eds.), *Interactive Whiteboards for Education: Theory, Research and Practice* (pp. 118–130). Hershey, PA: Information Science Reference.

Miller, T. K. (2012). A model for asynchronous discussions in a mathematics content course. In Juan, A., Huertas, M., Trenholm, S., & Steegmann, C. (Eds.), *Teaching Mathematics Online: Emergent Technologies and Methodologies* (pp. 1–21). Hershey, PA: Information Science Reference.

Miller, T. K. (2012). A theoretical framework for implementing technology for mathematics learning. In Ronau, R., Rakes, C., & Niess, M. (Eds.), *Educational Technology, Teacher Knowledge, and Classroom Impact: A Research Handbook on Frameworks and Approaches* (pp. 251–270). Hershey, PA: Information Science Publishing.

Mogollon, M. (2008). Number theory and finite fields. In Mogollon, M. (Ed.), *Cryptography and Security Services: Mechanisms and Applications* (pp. 33–50). Hershey, PA: CyberTech Publishing. doi:10.4018/978-1-59904-837-6.ch003.

Mokhasi, P., Adduci, J., & Kapadia, D. (2012). *Understanding differential equations using mathematica and interactive demonstrations*. Retrieved from http://www.codee.org/ref/CJ12-3093

Monaghan, J. (2004). Teachers' activities in technology-based mathematics lessons. *International Journal of Computers for Mathematical Learning*, *9*, 327–357. doi:10.1007/s10758-004-3467-6.

Moore, R. (1993). Does writing about science improve learning about science? *Journal of College Science Teaching*, *22*(4), 212–217.

Morge, S. P. (2009). Modeling in the classroom using squeak etoys. In Gibson, D., & Baek, Y. (Eds.), *Digital Simulations for Improving Education: Learning Through Artificial Teaching Environments* (pp. 334–349). Hershey, PA: Information Science Reference. doi:10.4018/978-1-60566-322-7.ch017.

Morrow, J., & Holland, J. (2008). Pask and ma join forces in an elementary mathematics methods course. In Luppicini, R. (Ed.), *Handbook of Conversation Design for Instructional Applications* (pp. 252–263). Hershey, PA: Information Science Reference. doi:10.4018/978-1-59904-597-9.ch016.

Morrow, J., & Holland, J. (2011). Pask and ma join forces in an elementary mathematics methods course. In Association, I. (Ed.), *Instructional Design: Concepts, Methodologies, Tools and Applications* (pp. 1806–1816). Hershey, PA: Information Science Reference. doi:10.4018/978-1-60960-503-2.ch718.

Nackaerts, O. (2010). Imaging in periodontology: 2D versus 3D visualization techniques. In Daskalaki, A. (Ed.), *Informatics in Oral Medicine: Advanced Techniques in Clinical and Diagnostic Technologies* (pp. 204–236). Hershey, PA: Medical Information Science Reference. doi:10.4018/978-1-60566-733-1.ch013.

Nason, R., & Woodruff, E. (2008). Innovations for online collaborative learning in mathematics. In Tomei, L. (Ed.), *Online and Distance Learning: Concepts, Methodologies, Tools, and Applications* (pp. 875–896). Hershey, PA: Information Science Reference. doi:10.4018/978-1-60566-026-4.ch323.

Nason, R., & Woodruff, E. (2009). Innovations for online collaborative learning in mathematics. In Khosrow-Pour, M. (Ed.), *Encyclopedia of Information Science and Technology* (2nd ed., pp. 2055–2060). Hershey, PA: Information Science Reference.

Nason, R., & Woodruff, E. (2009). Supporting online collaborative learning in mathematics. In Rogers, P., Berg, G., Boettcher, J., Howard, C., Justice, L., & Schenk, K. (Eds.), *Encyclopedia of Distance Learning* (2nd ed., pp. 1987–1993). Hershey, PA: Information Science Reference. doi:10.4018/978-1-60566-198-8.ch293.

National Council of Teachers of Mathematics (NCTM). (2000). *Principle and standards for school mathematics*. Reston, VA: NCTM.

Nayyar, V., Alam, M. Z., Kaminski, J., Kinnander, A., & Nyborg, L. (2013). An experimental investigation of the influence of cutting-edge geometry on the machinability of compacted graphite iron. *International Journal of Manufacturing, Materials, and Mechanical Engineering*, *3*(1), 1–25. doi:10.4018/ijmmme.2013010101.

Niess, M. L. (2012). Re-thinking pre-service mathematics teachers preparation: Developing technological, pedagogical, and content knowledge (TPACK). In Polly, D., Mims, C., & Persichitte, K. (Eds.), *Developing Technology-Rich Teacher Education Programs: Key Issues* (pp. 316–336). Hershey, PA: Information Science Reference. doi:10.4018/978-1-4666-0014-0.ch021.

Nissan, E. (2009). Argument structure models and visualization. In Pagani, M. (Ed.), *Encyclopedia of Multimedia Technology and Networking* (2nd ed., pp. 75–82). Hershey, PA: Information Science Reference.

Related References

Olaosun, I. E. (2010). Visualization of communication in some mobile phone directory and call log icons. In Taiwo, R. (Ed.), *Handbook of Research on Discourse Behavior and Digital Communication: Language Structures and Social Interaction* (pp. 490–496). Hershey, PA: Information Science Reference. doi:10.4018/978-1-61520-773-2.ch031.

Olaosun, I. E. (2012). Visualization of communication in some mobile phone directory and call log icons. In I. Management Association (Ed.), Computer Engineering: Concepts, Methodologies, Tools and Applications (pp. 940-947). Hershey, PA: Engineering Science Reference. doi: doi:10.4018/978-1-61350-456-7.ch410.

Oldknow, A., & Knights, C. (2013). *Mathematics education with digital technology*. London: Continuum International Publishing Group.

Oldknow, A., & Taylor, R. (2010). *Teaching mathematics using ICT (integrating information and communications technology in education)*. London: Continuum International Publishing Group.

Orrill, C. H., & Polly, D. (2012). Technology integration in mathematics: A model for integrating technology through content development. In Polly, D., Mims, C., & Persichitte, K. (Eds.), *Developing Technology-Rich Teacher Education Programs: Key Issues* (pp. 337–356). Hershey, PA: Information Science Reference. doi:10.4018/978-1-4666-0014-0.ch022.

Ouwehand, K., van Gog, T., & Paas, F. (2013). The use of gesturing to facilitate older adults' learning from computer-based dynamic visualizations. In Zheng, R., Hill, R., & Gardner, M. (Eds.), *Engaging Older Adults with Modern Technology: Internet Use and Information Access Needs* (pp. 33–58). Hershey, PA: Information Science Reference.

Owens, E., Song, H., & Kidd, T. T. (2007). Re-examining the socioeconomic factors affecting technology use in mathematics classroom practices. *International Journal of Web-Based Learning and Teaching Technologies, 2*(4), 72–87. doi:10.4018/jwltt.2007100105.

Owens, E., Song, H., & Kidd, T. T. (2009). Re-examining the socioeconomic factors affecting technology use in mathematics classroom practices. In Ang, C., & Zaphiris, P. (Eds.), *Human Computer Interaction: Concepts, Methodologies, Tools, and Applications* (pp. 1468–1483). Hershey, PA: Information Science Reference.

Palhares, P., & Gomes, A. (2008). A virtual laboratory of mathematics education. In Putnik, G., & Cruz-Cunha, M. (Eds.), *Encyclopedia of Networked and Virtual Organizations* (pp. 1797–1804). Hershey, PA: Information Science Reference. doi:10.4018/978-1-59904-885-7.ch238.

Papaodysseus, C., Panagopoulos, M., Rousopoulos, P., Arabadjis, D., Panopoulou, F., & Zannos, S. et al. (2012). Automatic identification of the writer of ancient Greek inscriptions, employing methods of computer engineering and mathematics. In Papaodysseus, C. (Ed.), *Pattern Recognition and Signal Processing in Archaeometry: Mathematical and Computational Solutions for Archaeology* (pp. 212–262). Hershey, PA: Information Science Reference.

Paravati, G., Sanna, A., Lamberti, F., & Ciminiera, L. (2010). On quality of experience in remote visualization on mobile devices. *International Journal of Mobile Human Computer Interaction*, *2*(1), 1–20. doi:10.4018/jmhci.2010100601.

Paravati, G., Sanna, A., Lamberti, F., & Ciminiera, L. (2012). On quality of experience in remote visualization on mobile devices. In Lumsden, J. (Ed.), *Social and Organizational Impacts of Emerging Mobile Devices: Evaluating Use* (pp. 1–19). Hershey, PA: Information Science Reference. doi:10.4018/978-1-4666-0194-9.ch001.

Pareja-Flores, C., Urquiza-Fuentes, J., & Velázquez Iturbide, J. Á. (2009). Web-based algorithm and program visualization for education. In Khosrow-Pour, M. (Ed.), *Encyclopedia of Information Science and Technology* (2nd ed., pp. 4093–4098). Hershey, PA: Information Science Reference.

Patterson, D. C., Reiniger, R. L., & Robertson, A. (2013). Cross-reality math visualization: The SubQuan system dream realizations in immersive environments, augmented realities, and virtual worlds. In D'Agustino, S. (Ed.), *Immersive Environments, Augmented Realities, and Virtual Worlds: Assessing Future Trends in Education* (pp. 57–77). Hershey, PA: Information Science Reference.

Perdue, D. S. (2012). Best practices for hybrid mathematics courses. In Juan, A., Huertas, M., Trenholm, S., & Steegmann, C. (Eds.), *Teaching Mathematics Online: Emergent Technologies and Methodologies* (pp. 90–118). Hershey, PA: Information Science Reference.

Piu, A. (2011). Design of a simulation game for the learning of mathematics. In Piu, A., & Fregola, C. (Eds.), *Simulation and Gaming for Mathematical Education: Epistemology and Teaching Strategies* (pp. 112–130). Hershey, PA: Information Science Reference.

Piu, A. (2011). Simulation games for the learning and teaching of mathematics. In Piu, A., & Fregola, C. (Eds.), *Simulation and Gaming for Mathematical Education: Epistemology and Teaching Strategies* (pp. 47–56). Hershey, PA: Information Science Reference.

Piu, A. (2011). Simulation games for the learning and teaching of mathematics. In I. Management Association (Ed.), Gaming and Simulations: Concepts, Methodologies, Tools and Applications (pp. 240-247). Hershey, PA: Information Science Reference. doi: doi:10.4018/978-1-60960-195-9.ch114.

Polly, D. (2013). *Common core mathematics standards and implementing digital technologies*. Hershey, PA: IGI Global.

Polly, D. (2013). Designing and teaching an online elementary mathematics methods course: Promises, barriers, and implications. In Hartshorne, R., Heafner, T., & Petty, T. (Eds.), *Teacher Education Programs and Online Learning Tools: Innovations in Teacher Preparation* (pp. 335–356). Hershey, PA: Information Science Reference.

Pongracz, F. (2009). Visualization and modelling in dental implantology. In Daskalaki, A. (Ed.), *Dental Computing and Applications: Advanced Techniques for Clinical Dentistry* (pp. 159–169). Hershey, PA: Medical Information Science Reference. doi:10.4018/978-1-60566-292-3.ch011.

Related References

Pongracz, F. (2011). Visualization and modelling in dental implantology. In I. Management Association (Ed.), Clinical Technologies: Concepts, Methodologies, Tools and Applications (pp. 2143-2152). Hershey, PA: Medical Information Science Reference. doi: doi:10.4018/978-1-60960-561-2.ch809.

Popejoy, K., & Polly, D. (2012). Leveraging technology to develop pre-service teachers' TPACK in mathematics and science methods courses. In Chen, I., & McPheeters, D. (Eds.), *Cases on Educational Technology Integration in Urban Schools* (pp. 170–182). Hershey, PA: Information Science Reference.

Porro, A. M. (2013). Preferred learning tools in online mathematics classes: Digital college learners. In Bryan, V., & Wang, V. (Eds.), *Technology Use and Research Approaches for Community Education and Professional Development* (pp. 66–84). Hershey, PA: Information Science Reference.

Porzio, D. (1999). Effects of differing emphases in the use of multiple representations and technology on students' understanding of calculus students. *Focus on Learning Problems in Mathematics*, *21*, 1–29.

Powell, A., & Ray, B. (2012). Supporting mathematics for young children through technology. In Blake, S., Winsor, D., & Allen, L. (Eds.), *Child Development and the Use of Technology: Perspectives, Applications and Experiences* (pp. 146–168). Hershey, PA: Information Science Reference.

Presmeg, N. (1986). Visualization and mathematical giftedness. *Educational Studies in Mathematics*, *17*, 287–311. doi:10.1007/BF00305075.

Quesada, A., & Maxwell, M. (1994). The effects of using graphing calculators to enhance college students' performance in precalculus. *Educational Studies in Mathematics*, *27*, 205–215. doi:10.1007/BF01278922.

Rahimi, S., Rias, R. A., & Khorasani, E. S. (2012). An open-bisimilarity based automated verification tool for calculus family of process calculi. *International Journal of Software Science and Computational Intelligence*, *4*(1), 55–83. doi:10.4018/jssci.2012010103.

Rahman, R. A., Yusof, Y. M., & Baharun, S. (2012). Realizing desired learning outcomes in undergraduate mathematics. In Yusof, K., Azli, N., Kosnin, A., Yusof, S., & Yusof, Y. (Eds.), *Outcome-Based Science, Technology, Engineering, and Mathematics Education: Innovative Practices* (pp. 182–206). Hershey, PA: Information Science Reference. doi:10.4018/978-1-4666-1809-1.ch009.

Raje, R. R., Chinnasamy, S., Olson, A. M., & Hidgon, W. (2007). The applications and enhancement of LePUS for specifying design patterns. In Taibi, T. (Ed.), *Design Pattern Formalization Techniques* (pp. 236–257). Hershey, PA: Idea Group Publishing. doi:10.4018/978-1-59904-219-0.ch011.

Rao, M. N. (2010). Creative and visualization tools in context of design. In Silva, A., & Simoes, R. (Eds.), *Handbook of Research on Trends in Product Design and Development: Technological and Organizational Perspectives* (pp. 139–158). Hershey, PA: Business Science Reference. doi:10.4018/978-1-61520-617-9.ch008.

Rav, Y. (1999). Why do we prove theorems? *Philosophia Mathematica*, *7*(3), 5–41. doi:10.1093/philmat/7.1.5.

Recio, A. M., & Godino, J. D. (2001). Institutional and personal meanings of proof. *Educational Studies in Mathematics*, *48*(1), 83–99. doi:10.1023/A:1015553100103.

Rehm, F., Klawonn, F., & Kruse, R. (2009). Visualization of high-dimensional data with polar coordinates. In Wang, J. (Ed.), *Encyclopedia of Data Warehousing and Mining* (2nd ed., pp. 2062–2067). Hershey, PA: Information Science Reference.

Reyes, V. M., & Sheth, V. (2011). Visualization of protein 3D structures in 'double-centroid' reduced representation: Application to ligand binding site modeling and screening. In Liu, L., Wei, D., Li, Y., & Lei, H. (Eds.), *Handbook of Research on Computational and Systems Biology: Interdisciplinary Applications* (pp. 583–598). Hershey, PA: Medical Information Science Reference. doi:10.4018/978-1-60960-491-2.ch026.

Reyes, V. M., & Sheth, V. (2013). Visualization of protein 3D structures in 'double-centroid' reduced representation: Application to ligand binding site modeling and screening. In I. Management Association (Ed.), Bioinformatics: Concepts, Methodologies, Tools, and Applications (pp. 1158-1173). Hershey, PA: Medical Information Science Reference. doi: doi:10.4018/978-1-4666-3604-0.ch059.

Romberg, T. A., Carpneter, T. P., & Fennema, E. (1993). Preface. In Romberg, T. A., Fennema, E., & Carpenter, T. P. (Eds.), *Integrating research on the graphical representation of functions* (pp. ix–xi). Hillsadale, NJ: Erlbaum.

Rose, B. (1989). Writing and mathematics: Theory and practice. In Connolly, P., & Vilardi, T. (Eds.), *Writing to learn mathematics and science* (pp. 15–30). New York: Teachers College Press.

Rosen, D. (2011). Methods for the measurement and visualization of social networks in multi-user virtual worlds. In Daniel, B. (Ed.), *Handbook of Research on Methods and Techniques for Studying Virtual Communities: Paradigms and Phenomena* (pp. 530–541). Hershey, PA: Information Science Reference.

Rosli, Z. B., & Shahbodin, F. (2012). Preliminary analysis of using games in problem based learning in learning mathematics. In Lu, Z. (Ed.), *Learning with Mobile Technologies, Handheld Devices, and Smart Phones: Innovative Methods* (pp. 202–208). Hershey, PA: Information Science Reference. doi:10.4018/978-1-4666-0936-5.ch013.

Rossol, N., Cheng, I., Jamal, I., Berezowski, J., & Basu, A. (2011). A real-time 3D visualization framework for multimedia data management, simulation, and prediction: Case study in geospatial-temporal biomedical disease surveillance networks. *International Journal of Multimedia Data Engineering and Management*, *2*(2), 1–18. doi:10.4018/jmdem.2011040101.

Rossol, N., Cheng, I., Jamal, I., Berezowski, J., & Basu, A. (2013). A real-time 3D visualization framework for multimedia data management, simulation, and prediction: Case study in geospatial-temporal biomedical disease surveillance networks. In Chen, S., & Shyu, M. (Eds.), *Multimedia Data Engineering Applications and Processing* (pp. 244–260). Hershey, PA: Information Science Reference. doi:10.4018/978-1-4666-2940-0.ch013.

Rubeo, T. (2008). Technology and the standards-based mathematics classroom. In Tomei, L. (Ed.), *Encyclopedia of Information Technology Curriculum Integration* (pp. 863–866). Hershey, PA: Information Science Reference. doi:10.4018/978-1-59904-881-9.ch135.

Related References

Salih, A. (2011). Laser scanning confocal imaging of forensic samples and their 3D visualization. In Daskalaki, A. (Ed.), *Digital Forensics for the Health Sciences: Applications in Practice and Research* (pp. 13–28). Hershey, PA: Medical Information Science Reference. doi:10.4018/978-1-60960-483-7.ch002.

Sauter, V. L., Mudigonda, S., Subramanian, A., & Creely, R. (2011). Visualization-based decision support systems: An example of regional relationship data. *International Journal of Decision Support System Technology, 3*(1), 1–20. doi:10.4018/jdsst.2011010101.

Sauter, V. L., Mudigonda, S., Subramanian, A., & Creely, R. (2013). Visualization-based decision support systems: an example of regional relationship data. In Power, D. (Ed.), *Engineering Effective Decision Support Technologies: New Models and Applications* (pp. 162–180). Hershey, PA: Engineering Science Reference.

Scheiter, K., Wiebe, E., & Holsanova, J. (2009). Theoretical and instructional aspects of learning with visualizations. In Zheng, R. (Ed.), *Cognitive Effects of Multimedia Learning* (pp. 67–88). Hershey, PA: Information Science Reference.

Scheiter, K., Wiebe, E., & Holsanova, J. (2011). Theoretical and instructional aspects of learning with visualizations. In Association, I. (Ed.), *Instructional Design: Concepts, Methodologies, Tools and Applications* (pp. 1667–1688). Hershey, PA: Information Science Reference. doi:10.4018/978-1-60960-503-2.ch710.

Schimke, D., Stoeger, H., & Ziegler, A. (2010). Evaluating the effectiveness of social visualization within virtual communities. In Dasgupta, S. (Ed.), *Social Computing: Concepts, Methodologies, Tools, and Applications* (pp. 783–801). Hershey, PA: Information Science Reference.

Selden, A., & Selden, J. (2002). Validations of proofs written as texts: Can undergraduates tell whether an argument proves a theorem? *Journal for Research in Mathematics Education, 34*(1), 4–36. doi:10.2307/30034698.

Shepard, R. G. (1993). Writing for conceptual development in mathematics. *The Journal of Mathematical Behavior, 12*, 287–293.

Shibli, A. (1992). Increasing learning with writing in quantitative and computer courses. *The Journal of Mathematical Behavior, 12*, 287–293.

Shiri, A. (2009). Metadata and metaphors in visual interfaces to digital libraries. In Theng, Y., Foo, S., Goh, D., & Na, J. (Eds.), *Handbook of Research on Digital Libraries: Design, Development, and Impact* (pp. 270–277). Hershey, PA: Information Science Reference. doi:10.4018/978-1-59904-879-6.ch027.

Shumar, W. (2009). Interaction, imagination and community building at the math forum. In Akoumianakis, D. (Ed.), *Virtual Community Practices and Social Interactive Media: Technology Lifecycle and Workflow Analysis* (pp. 269–282). Hershey, PA: Information Science Reference. doi:10.4018/978-1-60566-340-1.ch014.

Siau, K., Nah, F. F., Mennecke, B. E., & Schiller, S. Z. (2010). Co-creation and collaboration in a virtual world: A 3D visualization design project in second life. *Journal of Database Management, 21*(4), 1–13. doi:10.4018/jdm.2010100101.

Siau, K., Nah, F. F., Mennecke, B. E., & Schiller, S. Z. (2012). Co-creation and collaboration in a virtual world: A 3D visualization design project in second life. In Siau, K. (Ed.), *Cross-Disciplinary Models and Applications of Database Management: Advancing Approaches* (pp. 323–335). Hershey, PA: Information Science Reference.

Sierpinska, A. (1990). Some remarks on understanding in mathematics. *For the Learning of Mathematics*, *10*(3), 24–36.

Sikder, I. U. (2009). Rough sets and granular computing in geospatial information. In Karimi, H. (Ed.), *Handbook of Research on Geoinformatics* (pp. 154–159). Hershey, PA: Information Science Reference. doi:10.4018/978-1-59140-995-3.ch020.

Silva, R., Moura-Pires, J., & Santos, M. Y. (2012). Spatial clustering in SOLAP systems to enhance map visualization. *International Journal of Data Warehousing and Mining*, *8*(2), 23–43. doi:10.4018/jdwm.2012040102.

Simpson, E., & Butler, M. H. (2009). Analyzing communal tag relationships for enhanced navigation and user modeling. In Chevalier, M., Julien, C., & Soule-Dupuy, C. (Eds.), *Collaborative and Social Information Retrieval and Access: Techniques for Improved User Modeling* (pp. 43–64). Hershey, PA: Information Science Reference. doi:10.4018/978-1-60566-306-7.ch003.

Sindhu, M. S. (2009). New design approaches and a comparative study of taps packages for engineering education. *International Journal of Information and Communication Technology Education*, *5*(1), 38–52. doi:10.4018/jicte.2009010103.

Smith, A. M., Gentry, A. R., & Blake, S. (2012). Mathematics learning through the use of technology. In Blake, S., Winsor, D., & Allen, L. (Eds.), *Technology and Young Children: Bridging the Communication-Generation Gap* (pp. 199–221). Hershey, PA: Information Science Reference.

Smith, G. J. (2009). Information visualization and interface culture. In Braman, J., Vincenti, G., & Trajkovski, G. (Eds.), *Handbook of Research on Computational Arts and Creative Informatics* (pp. 196–212). Hershey, PA: Information Science Reference. doi:10.4018/978-1-60566-352-4.ch012.

Soundarajan, N., & Hallstrom, J. O. (2007). Precision, flexibility, and tool support: Essential elements of pattern formalization. In Taibi, T. (Ed.), *Design Pattern Formalization Techniques* (pp. 280–301). Hershey, PA: Idea Group Publishing. doi:10.4018/978-1-59904-219-0.ch013.

Stoecklin-Serino, C., & Paradice, D. (2009). An examination of the impacts of brand equity, security, and personalization on trust processes in an e commerce environment. *Journal of Organizational and End User Computing*, *21*(1), 1–36. doi:10.4018/joeuc.2009010101.

Tall, D. (1991). Intuition and rigor: The role of visualization in the calculus. In Zimmermann, W., & Cunningham, S. (Eds.), *Visualization in Teaching and Learning Mathematics* (pp. 105–119). Washington, DC: Mathematical Association of America.

Tall, D., & Shlomo, V. (1981). Concept image and concept definition in mathematics with particular reference to limits and continuity. *Educational Studies in Mathematics*, *12*, 151–169. doi:10.1007/BF00305619.

Tam, G. K., Lau, R. W., & Zhao, J. (2009). A 3D geometry model search engine to support learning. *International Journal of Distance Education Technologies, 7*(2), 100–112. doi:10.4018/jdet.2009040106.

Tam, G. K., Lau, R. W., & Zhao, J. (2011). A 3D geometry model search engine to support learning. In Jin, Q. (Ed.), *Distance Education Environments and Emerging Software Systems: New Technologies* (pp. 104–114). Hershey, PA: Information Science Reference. doi:10.4018/978-1-60960-539-1.ch007.

Tan, X., & Wang, Y. (2008). A denotational semantics of real-time process algebra (RTPA). *International Journal of Cognitive Informatics and Natural Intelligence, 2*(3), 57–70. doi:10.4018/jcini.2008070105.

Tempelaar, D. T., Rienties, B., Kaper, W., Giesbers, B., Schim van der Loeff, S., & van Gastel, L. … Cuypers, H. (2012). Mathematics bridging education using an online, adaptive e-tutorial: Preparing international students for higher education. In A. Juan, M. Huertas, S. Trenholm, & C. Steegmann (Eds.), Teaching Mathematics Online: Emergent Technologies and Methodologies(pp. 167-186). Hershey, PA: Information Science Reference. doi: doi:10.4018/978-1-60960-875-0.ch008.

Terashma, T., Makanae, K., & Dawood, N. (2011). Development of a 4D visualization tool for construction planning. In Matsuo, T., & Fujimoto, T. (Eds.), *E-Activity and Intelligent Web Construction: Effects of Social Design* (pp. 13–21). Hershey, PA: Information Science Reference. doi:10.4018/978-1-61520-871-5.ch002.

Tobias, S., Rudy, D., & Ispa, J. (2011). Relations between videogame play and 8th-graders' mathematics achievement. *International Journal of Gaming and Computer-Mediated Simulations, 3*(4), 33–53. doi:10.4018/jgcms.2011100103.

Tobias, S., Rudy, D., & Ispa, J. (2013). Relations between videogame play and 8th-graders' mathematics achievement. In Ferdig, R. (Ed.), *Design, Utilization, and Analysis of Simulations and Game-Based Educational Worlds* (pp. 240–261). Hershey, PA: Information Science Reference. doi:10.4018/978-1-4666-4018-4.ch015.

Tokmak, H. S., & Incikabi, L. (2013). Integration of the computer games into early childhood education pre-service teachers' mathematics teaching. In Keengwe, J. (Ed.), *Research Perspectives and Best Practices in Educational Technology Integration* (pp. 178–196). Hershey, PA: Information Science Reference.

Torriero, A., & Scancarello, I. (2009). An e-learning project for a basic mathematics course at the university. In Cartelli, A., & Palma, M. (Eds.), *Encyclopedia of Information Communication Technology* (pp. 256–267). Hershey, PA: Information Science Reference.

Trenholm, S., Juan, A. A., Simosa, J., & Oliveira, A. (2012). Long-term experiences in mathematics e-learning in Europe and the USA. In Juan, A., Huertas, M., Trenholm, S., & Steegmann, C. (Eds.), *Teaching Mathematics Online: Emergent Technologies and Methodologies* (pp. 238–257). Hershey, PA: Information Science Reference.

Trinchero, R. (2013). PotenziaMente: A collection of online games for learning mathematics and enhance thought processes. In Pumilia-Gnarini, P., Favaron, E., Pacetti, E., Bishop, J., & Guerra, L. (Eds.), *Handbook of Research on Didactic Strategies and Technologies for Education: Incorporating Advancements* (pp. 68–77). Hershey, PA: Information Science Reference.

Uribe-Flórez, L. J., & Trespalacios, J. (2013). The role of animations and manipulatives in supporting learning and communication in mathematics classrooms. In Baek, Y., & Whitton, N. (Eds.), *Cases on Digital Game-Based Learning: Methods, Models, and Strategies* (pp. 400–410). Hershey, PA: Information Science Reference.

Ursyn, A. (2010). Aesthetic expectations for information visualization. *International Journal of Creative Interfaces and Computer Graphics, 1*(1), 19–39. doi:10.4018/jcicg.2010010103.

Ursyn, A. (2012). Aesthetic expectations for information visualization. In Falchuk, B., & Marcos, A. (Eds.), *Innovative Design and Creation of Visual Interfaces: Advancements and Trends* (pp. 13–33). Hershey, PA: Information Science Reference. doi:10.4018/978-1-4666-0285-4.ch002.

Van Tien, L., Tho, Q. T., & Cheung, H. S. (2011). Mathematical retrieval techniques for online mathematics learning. In Daniel, B. (Ed.), *Handbook of Research on Methods and Techniques for Studying Virtual Communities: Paradigms and Phenomena* (pp. 394–409). Hershey, PA: Information Science Reference.

Vande Moere, A., & Offenhuber, D. (2009). Beyond ambient display: A contextual taxonomy of alternative information display. *International Journal of Ambient Computing and Intelligence, 1*(2), 39–46. doi:10.4018/jaci.2009040105.

Viktor, H. L., & Paquet, E. (2009). Visual data mining from visualization to visual information mining. In Wang, J. (Ed.), *Encyclopedia of Data Warehousing and Mining* (2nd ed., pp. 2056–2061). Hershey, PA: Information Science Reference.

Vinner, S. (1989). The avoidance of visual consideration in calculus students. *Focus on Learning Problems in Mathematics, 11*, 149–156.

Vlieger, E., & Leydesdorff, L. (2012). Visualization and analysis of frames in collections of messages: Content analysis and the measurement of meaning. In Mora, M., Gelman, O., Steenkamp, A., & Raisinghani, M. (Eds.), *Research Methodologies, Innovations and Philosophies in Software Systems Engineering and Information Systems* (pp. 321–339). Hershey, PA: Information Science Reference. doi:10.4018/978-1-4666-0179-6.ch016.

Wallner, G., Kriglstein, S., & Biba, J. (2013). Evaluating games in classrooms: A case study with DOGeometry. In Baek, Y., & Whitton, N. (Eds.), *Cases on Digital Game-Based Learning: Methods, Models, and Strategies* (pp. 475–494). Hershey, PA: Information Science Reference.

Wang, J. (2013). Pedagogy and curriculum in architecture and engineering. In Challenging ICT Applications in Architecture, Engineering, and Industrial Design Education (pp. 65-92). Hershey, PA: Engineering Science Reference. doi: doi:10.4018/978-1-4666-1999-9.ch004.

Related References

Wang, J. (2013). Professionalism in architecture and engineering. In Challenging ICT Applications in Architecture, Engineering, and Industrial Design Education (pp. 137-156). Hershey, PA: Engineering Science Reference. doi: doi:10.4018/978-1-4666-1999-9.ch007.

Wang, Y. (2008). Deductive semantics of RTPA. *International Journal of Cognitive Informatics and Natural Intelligence*, *2*(2), 95–121. doi:10.4018/jcini.2008040106.

Wang, Y. (2008). Formal RTPA models for a set of meta-cognitive processes of the brain. *International Journal of Cognitive Informatics and Natural Intelligence*, *2*(4), 15–28. doi:10.4018/jcini.2008100102.

Wang, Y. (2008). On concept algebra: A denotational mathematical structure for knowledge and software modeling. *International Journal of Cognitive Informatics and Natural Intelligence*, *2*(2), 1–19. doi:10.4018/jcini.2008040101.

Wang, Y. (2008). On system algebra: A denotational mathematical structure for abstract system modeling. *International Journal of Cognitive Informatics and Natural Intelligence*, *2*(2), 20–43. doi:10.4018/jcini.2008040102.

Wang, Y. (2008). RTPA: A denotational mathematics for manipulating intelligent and computational behaviors. *International Journal of Cognitive Informatics and Natural Intelligence*, *2*(2), 44–62. doi:10.4018/jcini.2008040103.

Wang, Y. (2009). A cognitive informatics reference model of autonomous agent systems (AAS). *International Journal of Cognitive Informatics and Natural Intelligence*, *3*(1), 1–16. doi:10.4018/jcini.2009010101.

Wang, Y. (2009). A denotational mathematics for manipulating intelligent and computational behaviors. In Tiako, P. (Ed.), *Software Applications: Concepts, Methodologies, Tools, and Applications* (pp. 2943–2962). Hershey, PA: Information Science Reference. doi:10.4018/978-1-60566-060-8.ch172.

Wang, Y. (2009). Formal descriptions of cognitive processes of perceptions on spatiality, time, and motion. *International Journal of Cognitive Informatics and Natural Intelligence*, *3*(2), 84–98. doi:10.4018/jcini.2009040105.

Wang, Y. (2009). On abstract intelligence: Toward a unifying theory of natural, artificial, machinable, and computational intelligence. *International Journal of Software Science and Computational Intelligence*, *1*(1), 1–17. doi:10.4018/jssci.2009010101.

Wang, Y. (2009). The theoretical framework of cognitive informatics. In Wang, Y. (Ed.), *Novel Approaches in Cognitive Informatics and Natural Intelligence* (pp. 1–27). Hershey, PA: Information Science Reference.

Wang, Y. (2010). RTPA: A denotational mathematics for manipulating intelligent and computational behaviors. In Wang, Y. (Ed.), *Discoveries and Breakthroughs in Cognitive Informatics and Natural Intelligence* (pp. 178–199). Hershey, PA: Information Science Reference.

Wang, Y. (2011). Inference algebra (IA): A denotational mathematics for cognitive computing and machine reasoning (I). *International Journal of Cognitive Informatics and Natural Intelligence*, *5*(4), 61–82. doi:10.4018/jcini.2011100105.

Wang, Y. (2012). Inference algebra (IA): A denotational mathematics for cognitive computing and machine reasoning (II). *International Journal of Cognitive Informatics and Natural Intelligence*, *6*(1), 21–47. doi:10.4018/jcini.2012010102.

Wang, Y. (2013). Inference algebra (IA): A denotational mathematics for cognitive computing and machine reasoning (I). In Wang, Y. (Ed.), *Cognitive Informatics for Revealing Human Cognition: Knowledge Manipulations in Natural Intelligence* (pp. 159–177). Hershey, PA: Information Science Reference.

Wang, Y., & Patel, S. (2009). Exploring the cognitive foundations of software engineering. *International Journal of Software Science and Computational Intelligence*, *1*(2), 1–19. doi:10.4018/jssci.2009040101.

Wang, Y., Zadeh, L. A., & Yao, Y. (2009). On the system algebra foundations for granular computing. *International Journal of Software Science and Computational Intelligence*, *1*(1), 64–86. doi:10.4018/jssci.2009010105.

Warner, Z. B. (2011). Adoption of computer-based formative assessment in a high school mathematics classroom. *Journal of Cases on Information Technology*, *13*(4), 9–20. doi:10.4018/JCIT.2011100102.

Wasburn, M. H. (2008). Cultivating greater acceptance of women in technology: A pilot study. In Van Slyke, C. (Ed.), *Information Communication Technologies: Concepts, Methodologies, Tools, and Applications* (pp. 3552–3565). Hershey, PA: Information Science Reference. doi:10.4018/978-1-59904-949-6.ch251.

Washburn, M. H. (2009). Media and women in technology. In Tomei, L. (Ed.), *Information Communication Technologies for Enhanced Education and Learning: Advanced Applications and Developments* (pp. 1–14). Hershey, PA: Information Science Reference.

Weber, K. (2003). Students' difficulties with proof. *MAA On-line Research Sampler*. Retrieved from http://www.maa.org/

White, K. F., & Wasburn, M. H. (2006). CareerQuesting: Evaluating web-based resources for interesting girls in STEM careers. *International Journal of Information and Communication Technology Education*, *2*(3), 45–59. doi:10.4018/jicte.2006070104.

White, K. S., & Wasburn, M. H. (2008). CareerQuesting revisited: A protocol for increasing girls' interest in STEM careers. In Tomei, L. (Ed.), *Adapting Information and Communication Technologies for Effective Education* (pp. 121–134). Hershey, PA: Information Science Reference.

Xu, H. (2009). Consumer responses to the introduction of privacy protection measures: An exploratory research framework. *International Journal of E-Business Research*, *5*(2), 21–47. doi:10.4018/jebr.2009040102.

Xu, X. (2009). Feature technology. In Integrating Advanced Computer-Aided Design, Manufacturing, and Numerical Control: Principles and Implementations (pp. 75-89). Hershey, PA: Information Science Reference. doi: doi:10.4018/978-1-59904-714-0.ch004.

Xu, X. (2009). Geometric modelling and computer-aided design. In Integrating Advanced Computer-Aided Design, Manufacturing, and Numerical Control: Principles and Implementations (pp. 1-31). Hershey, PA: Information Science Reference. doi: doi:10.4018/978-1-59904-714-0.ch001.

Related References

Xu, Z., Li, W., Fu, H., & Zeng, Z. (2003). Mathematics education over the internet based on vega grid technology. *International Journal of Distance Education Technologies*, *1*(3), 1–13. doi:10.4018/jdet.2003070101.

Xu, Z., Li, W., Fu, H., & Zeng, Z. (2008). Mathematics education over the internet based on vega grid technology. In Tomei, L. (Ed.), *Online and Distance Learning: Concepts, Methodologies, Tools, and Applications* (pp. 1362–1372). Hershey, PA: Information Science Reference.

Yahia, S. B., Couturier, O., Hamrouni, T., & Nguifo, E. M. (2009). Meta-knowledge based approach for an interactive visualization of large amounts of association rules. In Zhao, Y., Zhang, C., & Cao, L. (Eds.), *Post-Mining of Association Rules: Techniques for Effective Knowledge Extraction* (pp. 200–223). Hershey, PA: Information Science Reference. doi:10.4018/978-1-60566-404-0.ch011.

Yamamoto, C. H., Ferreira de Oliveira, M. C., & Rezende, S. O. (2009). Visualization to assist the generation and exploration of association rules. In Zhao, Y., Zhang, C., & Cao, L. (Eds.), *Post-Mining of Association Rules: Techniques for Effective Knowledge Extraction* (pp. 224–245). Hershey, PA: Information Science Reference. doi:10.4018/978-1-60566-404-0.ch012.

Yau, S., Davulcu, H., Mukhopadhyay, S., Huang, D., Gong, H., Singh, P., & Gelgi, F. (2007). Automated situation-aware service composition in service-oriented computing. *International Journal of Web Services Research*, *4*(4), 59–82. doi:10.4018/jwsr.2007100103.

Yoong, S., & Lew, L. Y. (2010). Enhancing ICT application in science and mathematics education: The Malaysian smart school experience. In Rodrigues, S. (Ed.), *Multiple Literacy and Science Education: ICTs in Formal and Informal Learning Environments* (pp. 142–164). Hershey, PA: Information Science Reference.

Yu, K., Zhang, N., Xue, G., & Yao, S. (2013). Advanced branching and synchronization patterns description using pi-calculus. In Lu, Z. (Ed.), *Design, Performance, and Analysis of Innovative Information Retrieval* (pp. 394–405). Hershey, PA: Information Science Reference.

Yusof, K. M., Azli, N. A., Kosnin, A. M., Yusof, S. K., & Yusof, Y. M. (2012). *Outcome-based science, technology, engineering, and mathematics education: Innovative practices*. Hershey, PA: IGI Global. doi:10.4018/978-1-4666-1809-1.

Zandieh, M. (1999). The role of a formal definition in nine students' concept image of derivative. In Berenson, Dawkins, Blanton, Coulombe, Kolb, Norwood, & Stiff (Eds.), *Proceedings of the 20thConference of the North American Chapter of the International Group for the Psychology of Mathematics Education*, (vol. 1, pp. 136-141). Psychology of Mathematics Education.

Zelinka, I., Davendra, D., Jašek, R., & Šenkeřík, R. (2012). Investigation on visualization, analysis, and control of complex networks dynamics. *International Journal of Energy Optimization and Engineering*, *1*(3), 48–73. doi:10.4018/ijeoe.2012070103.

Zhu, H., Wang, L., & Chan, T. K. (2009). Parallel and distributed visualization advances. In Khosrow-Pour, M. (Ed.), *Encyclopedia of Information Science and Technology* (2nd ed., pp. 3018–3025). Hershey, PA: Information Science Reference.

Zimmermann, W. (1991). Visual thinking in calculus. In Zimmermann, W., & Cunningham, S. (Eds.), *Visualization in teaching and learning mathematics* (pp. 127–137). Washington, DC: Mathematical Association of America.

Zinger, S., Morvan, Y., Ruijters, D., Do, L., & de With, P. H. (2012). Multi-view autostereoscopic visualization using bandwidth-limited channels. In Malik, A., Choi, T., & Nisar, H. (Eds.), *Depth Map and 3D Imaging Applications: Algorithms and Technologies* (pp. 363–378). Hershey, PA: Information Science Reference.

Compilation of References

Abboud-Blanchard, M. (2005). Uses of ICT by pre-service teachers. In F. Olivero & R. Sutherland (Eds.), *Proceedings of the 7th International Conference of Technology in Mathematics Teaching*, (Vol. 2, pp. 74 - 78). Technology in Mathematics Teaching.

Albers, D. J., Anderson, R. D., & Loftsgaarden, D. O. (Eds.). (1986). Undergraduate programs in the mathematics and computer sciences: The 1985-1986 survey. Mathematical Association of America, 7.

Arcavi, A. (1999). The role of visual representations in the learning of mathematics. In F. Hitt & M. Santos (Eds.), *Proceedings of the 21st North American PME Conference*, (vol. 1, pp. 55-80). PME.

Arcavi, A. (2008). Modeling with graphical representations. *For the Learning of Mathematics*, *28*(2), 2–10.

Architect, O. D. E. (2010). *Consortium for ordinary differential equations*. New York: John Wiley & Sons, Inc..

Artigue, M. (2002). Learning mathematics in a CAS environment: The genesis of a reflection about instrumentation and the dialectics between technical and conceptual work. *International Journal of Computers for Mathematical Learning*, *7*, 245–274. doi:10.1023/A:1022103903080.

Arzarello, F., Micheletti, C., Olivero, F., Robutti, O., Paola, D., & Gallino, G. (1998). A model for analysing the transition to formal proofs in geometry. In *Proceedings of PME 22*. Stellenbosch, South Africa: PME Press.

Ascher, U. M., & Greif, C. (2011). *A first course in numerical methods*. Philadelphia, PA: Society for Industrial and Applied Mathematics. doi:10.1137/1.9780898719987.

Austin, A. K. (1971). Problems and solutions. *Mathematics Magazine*, *44*(1), 16–19.

Autograph Software. (n.d.). Retrieved from www.autograph-maths.com

Autograph Specific Resources. (n.d.). Retrieved from www.tsm-resources.com/autograph

Bakker, A. (2004). *Design research in statistics education: On symbolizing and computer tools*. (Doctoral dissertation). Utrecht University, Utrecht, The Netherlands.

Bakker, A. (2003). Reasoning about shape as a pattern in variability. In Lee, C. (Ed.), *Reasoning about Variability: A Collection of Current Research Studies*. Dordrecht, The Netherlands: Kluwer Academic Publisher.

Barnsley, M. (2000). *Fractals everywhere* (2nd ed.). San Francisco, CA: Morgan Kaufman.

Beckmann, A. (2001). Probleme beim beweisenlernen–DGS als lösung? In Elschenbroich, Gawlick, & Henn (eds.), Zeichnung–Figur–Zugfigur. Hildesheim, Germany: Franzbecker.

Ben-Zvi, D. (2006). Scaffolding students' informal inference and argumentation. In A. Rossman & B. Chance (Eds.), *Working Cooperatively in Statistics Education: Proceedings of the Seventh International Conference of Teaching Statistics*. Salvador, Brazil: ICOTS.

Ben-Zvi, D. (2000). Toward understanding the role of technological tools in statistical learning. *Mathematical Thinking and Learning*, *2*, 127–155. doi:10.1207/S15327833MTL0202_6.

Biehler, R. (1995). *Toward requirements for more adequate software tools that support both: Learning and doing statistics*. Bielefeld, Germany: University of Bielefeld.

Blanchard, P., Devaney, R. L., & Hall, G. R. (2011). *Differential equations* (4th ed.). London: Brooks Cole.

Blyth, B., & Labovic, A. (2009). Using maple to implement elearning integrated with computer and assessment. *International Journal of Mathematical Education in Science and Technology*, *40*(7), 975–988. doi:10.1080/00207390903226856.

Boaler, J. (2002). The development of disciplinary relationships: Knowledge, practice and identity in mathematics classrooms. *For the Learning of Mathematics*, *22*(1), 42–47.

Boero, P. (1999). Argumentation and mathematical proof: A complex, productive, unavoidable relationship in mathematics and mathematics education. *International Newsletter on the Teaching and Learning of Mathematical Proof, 7/8*.

Boole, M. E. (1904). *The preparation of the child for science*. Clarendon Press.

Boole, M. E. (1909). *Philosophy and fun of algebra*. C. W. Daniel.

Borges, J. (1975). *El libro de arena*. Madrid, Spain: Alianza Editorial S.A..

Borrelli, R., & Coleman, C. (2004). *Differential equations: A modeling perspective*. New York: Wiley.

Bowker, G. C., & Star, S. L. (1999). *Sorting things out: Classification and its consequences*. Cambridge, MA: MIT Press.

Brockmann-Behnsen, D. (2011). A long-term educational treatment using dynamic geometry software. In *Proceedings of ICTMT 10*. Portsmouth, England: Open University Press.

Burton, L. (1999). The practices of mathematicians: What do they tell us about coming to know mathematics? *Educational Studies in Mathematics*, *31*, 127–143.

Butler, D. (n.d.). *Blog*. Retrieved from http://dougbutler.wordpress.com/

Cain, J. W. (2011). Taking math to heart: Mathematical challenges in cardiac electrophysiology. *Notices of the American Mathematical Society*, *58*(4), 542–549.

Clarke, M. (2001). *The book of sand - A hypertext/puzzle*. Retrieved July 8, 2011, from http://artificeeternity.com/bookofsand/

Cobb, P., & Yackel, E. (1996). Constructivist, emergent, and sociocultural perspectives in the context of developmental research. *Educational Psychologist*, *31*(3/4), 175–190.

C-ODE-E (NSF Consortium for Ordinary Differential Equations Experiments). (1998). *ODE architect*. New York: Wiley.

Danby, J. M. A. (1985). *Computing applications to differential equations: Modelling in the physical and social sciences*. Reston, VA: Reston Publishing Co.

Danby, J. M. A. (1997). *Computer modeling: From sports to spaceflight... from order to chaos*. Richmond, VA: Willman Bell, Inc.

Deinhardt, A. (2000). Die ortskurven des höhenschnittpunktes. *Vechtaer Fachdidaktische Forschungen und Berichte, 3*.

Dienes, Z. (1960). *Building up mathematics* (4th ed.). London, UK: Hutchinson Educational Ltd..

Diković. (2009). Applications GeoGebra into teaching some topics of mathematics at the college level. *The International Journal of Computer Science and Information Systems*, *6*(2).

Diković. Ivković, & Milivojević, M. (2010). *Visualizing and understanding limits via internet technologies*. Paper presented at the 3rd International Conference Science and Higher Education in Function of Sustainable Development. Užice, Srbija.

Duval, R. (1999). Representation, vision and visualization: Cognitive functions in mathematical thinking – Basic issues for learning. In F. Hitt & M. Santos (Eds.), *Proceedings of the 21st North American PME Conference*, (vol. 1, pp. 3-26). PME.

Dynamic Online Content of this Chapter. (n.d.). Retrieved from www.tsm-resources.com/dynamic

Eisenberg, T., & Dreyfus, T. (1991). On visual versus analytical thinking in mathematics. I C. Hoyles, R. Noss, & R. Sutherland (Eds.), *Proceedings of the Tenth International Conference on the Psychology of Mathematics Education*, (vol. 1, pp. 153-158). Psychology of Mathematics Education.

Elschenbroich, H.-J. (1997). Tod des beweisens oder wiederauferstehung? In Hischer, H. (Ed.), *Computer and Geometrie. Neue Chancen für den Geometrieunterricht?* Hildesheim, Germany: Franzbecker.

Elschenbroich, H.-J. (2001). *Electronic interactive worksheets–A visual and dynamic way of learning geometry*. Montreal: Cabri World.

Elschenbroich, H.-J., & Seebach, G. (2002). *Dynamisch geometrie entdecken. Klasse 7, 8, 9, 10*. Rosenheim, Germany: CoTec.

Erickson, T. (2006). Using simulation to learn about inference. In A. Rossman & B. Chance (Eds.), *Working Cooperatively in Statistics Education: Proceedings of the Seventh International Conference of Teaching Statistics* (ICOTS-7). Salvador, Brazil: ICOTS.

Erickson, T. (2001). *Data in depth: Exploring mathematics with fathom*. Emeryville, CA: Key Curriculum Press.

Farlow, J., Hall, J. E., McDill, J. M., & West, B. H. (2007). *Differential equations and linear algebra* (2nd ed.). New York: Pearson.

Finzer, W. (2000). *Design of fathom: A dynamic statistics environment, for the teaching of mathematics*. Retrieved July 2, 2011 from http://www.keypress.com/Documents/fathom/Bibliography/Fathom_in_Teaching_Math.pdf

Finzer, W., & Erickson, T. (1998). DataSpace — A computer learning environment for data analysis and statistics based on dynamic dragging, visualization, simulation, and networked collaboration. In *Proceedings of the Fifth International Conference on Teaching of Statistics* (pp. 825-830). Voorburg, The Netherlands: International Statistical Institute.

Finzer, W. (2001). *Fathom*. Emeryville, CA: Key Curriculum Press.

Forehand, G. (1967). *Individual differences in problem solving processes of college students*. Pittsburgh, PA: Carnegie Mellon University.

Franklin, C. A., Kader, G., Mewborn, D., Moreno, J., Peck, R., Perry, M., & Scheaffer, R. (2007). *Guidelines for assessment and instruction in statistics education (GAISE) report: A pre-K–12 curriculum framework*. Alexandria, VA: American Statistical Association.

Freudenthal, H. (1973). *Mathematics as an educational task*. Dordrecht, Netherlands: Reidel.

Garafalo, J., Drier, H. S., Harper, S., & Timmerman, M. A. (2000). Promoting appropriate uses of technology in mathematics teacher preparation. *Contemporary Issues in Technology & Teacher Education, 1*(1). Retrieved from http://www.citejournal.org/vol1/iss1/currentissues/mathematics/article1.htm.

Gardner, H., & Hudson, I. (1999). University students' ability to apply statistical procedures. *Journal of Statistics Education, 7*(1). Retrieved July 2, 2011 from www.amstat.org/publications/jse/secure/v7n1/gardner.cfm

Garfield, J., & Ahlgren, A. (1998). Difficulties in learning basic concepts in probability and statistics: Implications for research. *Journal for Research in Mathematics Education, 19*, 44–63. doi:10.2307/749110.

Gawlick, T. (2002). On dynamic geometry software in the regular classroom. *Zentralblatt für didaktik der mathematic, 34*(3), 85-92.

Gawlick, T. (2003). Utilizing DGS as heuristic tool in introductory calculus courses. In *Proceedings of ICTME 2*. Beirut, Lebanon: ICTME Press.

Gawlick, T. (2005). Connecting arguments to actions–Dynamic geometry as means for the attainment of higher van hiele levels. *ZDM. Zentralblatt für Didaktik der Mathematik, 37*(5).

Gil, E., & Ben-Zvi, D. (2011). Explanations and context in the emergence of students' informal inferential reasoning. *Mathematical Thinking and Learning, 13*(1), 87–108. doi:10.1080/10986065.2011.538295.

Goos, M. (2004). Learning mathematics in a classroom community of inquiry. *Journal for Research in Mathematics Education, 35*, 258–291. doi:10.2307/30034810.

Gordon, F. S., & Gordon, S. P. (1992). Sampling + simulation = statistical understanding. In Gordon, F. S. (Ed.), *Statistics for the Twenty-First Century* (pp. 207–216). Washington, DC: The Mathematical Association of America.

Gottlieb, H. P. W. (1996). What is the simplest jerk function that gives chaos? *American Journal of Physics, 64*, 525. doi:10.1119/1.18276.

Graham, A., & Thomas, M. (2000). Building a versatile understanding of algebraic variables with a graphic calculator. *Educational Studies in Mathematics, 41*(3), 265–282. doi:10.1023/A:1004094013054.

Gravemeijer, K. (1994). *Developing realistic mathematics education.* Utrecht, The Netherlands: CD-β Press.

Green, D. G. (1982). A survey of probability concepts in 3000 students aged 11-16. In D. V. Grey (Ed.), *Proceedings of the First International Conference on Teaching Statistics* (pp. 766-783). London: Statistics Teaching Trust.

Habre, S., & McDill, J. (2008). A complete classification of 2×2 linear iterative systems. *The Electronic Journal of Mathematics and Technology, 2*(1).

Habre, S. (2000). Exploring students' strategies to solve ordinary differential equations in a reformed setting. *The Journal of Mathematical Behavior, 18*, 455–472. doi:10.1016/S0732-3123(00)00024-9.

Habre, S. S., Hubbard, J. H., & West, B. H. (2001). The convergence of an Euler approximation of an initial value problem is not always obvious. *The American Mathematical Monthly, 108*(4), 326–335. doi:10.2307/2695239.

Hadley, J., & Singmaster, D. (1992). Problems to sharpen the young. *The Mathematical Gazette, 76*(475), 102–126. doi:10.2307/3620384.

Hammerman, J., & Rubin, A. (2003). Reasoning in the presence of variability. In Lee, C. (Ed.), *Reasoning about Variability: A Collection of Current Research Studies.* Dordrecht, The Netherlands: Kluwer Academic Publisher.

Harvey Mudd College. (2011). *ODE toolkit.* Retrieved from http://odetoolkit.hmc.edu/

Hayes, J. (1966). Memory, goals, and problem solving. In Kleinmuntz, B. (Ed.), *Problem Solving: Research, Method, and Theory.* New York: Wiley.

Haynes, M., & Upton, D. S. (2008). Computer manipulatives in an ordinary differential equations course: Development, implementation, and assessment. *Journal of Science Education and Technology, 17*, 124–137. doi:10.1007/s10956-007-9058-2.

Healy, L., Holzl, R., Hoyles, C., & Noss, R. (1994). Messing up. *Micromath, 10*(1), 14–16.

Heath, M. (2002). *Scientific computing: An introductory survey* (2nd ed.). New York: McGraw-Hill.

Hefendehl-Hebeker, L., & Hölzl, R. (2000). Zerlegbare dreiecke im problemorientierten unterricht. In Baptist (ed.), Mathematikunterricht im Wandel. Bamberg, Germany: Buchner.

Heller, J., Curtis, Jaffer, & Verboncoeur. (2005). *Impact of handheld graphing calculator use on student achievement in algebra 1.* Heller Research Associates.

Hemmings, R., & Tahta, D. (1992). *Images of infinity.* London: Tarquin Publications.

Herzig, A. (2004). Becoming mathematicians: Women and students of color choosing and leaving doctoral mathematics. *Review of Educational Research, 74*(2), 171–214. doi:10.3102/00346543074002171.

Hohenwarter, M., Hohenwarter, J., Kreis, Y., & Lavicza, Z. (2008). Teaching and learning calculus with free dynamic mathematics software GeoGebra. In *Proceedings of the 11th International Congress on Mathematics Education.* University of Nuevo Leon.

Hohn, H. (2005). *Iterated linear systems,* unpublished.

Hohn, H., West, B. H., Strogatz, S. H., McDill, J. M., & Cantwell, J. (2000). *Interactive differential equations.* Reading, MA: Addison Wesley. Retrieved from http://www.aw-bc.com/ide

Hölzl, R. (1996). How does 'dragging' affect the learning of geometry? *International Journal of Computers for Mathematical Learning, 1*(2), 169–187. doi:10.1007/BF00571077.

Hölzl, R. (1999). *Qualitative unterrichtsstudien zur verwendung dynamischer geometriesoftware.* Augsburg, Germany: Wissner.

Hölzl, R. (2001). Using DGS to add contrast to geometric situations. *International Journal of Computers for Mathematical Learning, 6*(1), 63–86. doi:10.1023/A:1011464425023.

Hoyles, C., & Noss, R. (1994). Dynamic geometry environments: What is the point? *Mathematics Teacher, 87,* 716–717.

Hoyles, C., Noss, R., Kent, P., & Bakker, A. (2010). *Improving mathematics at work: The need for techno-mathematical literacies.* New York: Routledge.

Hubbard, J. (1994). What it means to understand a differential equation. *The College Mathematics Journal, 25*(5), 372–384.

Hubbard, J. H. (1999). The forced damped pendulum: chaos, complication and control. *The American Mathematical Monthly, 106,* 741–758. doi:10.2307/2589021.

Hubbard, J. H. (2008). *The forced damped pendulum: Chaos, complication and control.* Retrieved from http://www.math.cornell.edu/~hubbard/Wisconsinpendulumtalk.pdf

Hubbard, J. H., & Lundell, B. E. (2011, March). A first look at differential algebra. *The American Mathematical Monthly,* 245–261.

Hubbard, J. H., McDill, J. M., Noonburg, A., & West, B. H. (1994). A new look at the airy equation with fences and funnels. *The College Mathematics Journal, 25*(5), 419–431.

Hubbard, J.H., & West, B. H. (1991). Differential equations: A dynamical systems approach: Part I: Ordinary differential equations. *Texts in Applied Mathematics,* (5).

Hubbard, J. H., & West, B. H. (1992). *MacMath 9.2.* New York: Springer Verlag. Hubbard, J.H., & West, B.H. (1995). Differential equations: A dynamical systems approach: Part II: Higher-dimensional systems. *Texts in Applied Mathematics,* (18).

Hubbard, J. H., & West, B. H. (1993). *MacMath.* New York: Springer-Verlag.

Hubbard, J. H., & West, B. H. (1995). *Differential equations: A dynamical systems approach, part II: Higher dimensional systems.* New York: Springer Verlag.

Hubbard, J. H., & West, B. H. (1999). *MacMath 10.0.* Unpublished.

IES. (n.d.). Retrieved from http://www.ies.co.jp/math/java/calc/doukan/doukan.html

Kaput, J. (1992). Technology and mathematics education. In Grouws, D. A. (Ed.), *Handbook of Research on Mathematics Teaching and Learning* (pp. 515–556). New York: Macmillan.

Kellogg, M., & Kersaint, G. (2004). Creating a vision for the Standards using online videos in an elementary mathematics methods course. *Contemporary Issues in Technology and Teacher, Education, 4*(1). Retrieved October 14, 2007, from http://www.citejournal.org/vol4/iss1/mathematics/article1.cfm

Kersaint, G. (2003). Technology beliefs and practices of mathematics education faculty. *Journal of Technology and Teacher Education, 11*(4), 549–577. Retrieved from http://www.editlib.org/p/2222.

Kersaint, G. (2007). Toward technology integration in mathematics education: A technology-integration course planning assignment. *Contemporary Issues in Technology & Teacher Education, 7*(4). Retrieved from http://www.citejournal.org/vol7/iss4/mathematics/article1.cfm.

Kilpatrick, J. (1967). *Analyzing the Solution of Word Problems in Mathematics: An Exploratory Study.* (Dissertation). Stanford, CA, Stanford University.

Klein, F. (1924). *Elementarmathematik vom höheren standpunkte aus.* Berlin: Springer. doi:10.1007/978-3-642-90853-8.

Konold, C. (2002). Teaching concepts rather than conventions. *New England Journal of Mathematics, 34*(2), 69–81.

Konold, C., & Miller, C. D. (2005). *TinkerPlots: Dynamic data exploration.* Emeryville, CA: Key Curriculum Press.

Kwon, O. N., Rasmussen, C., & Allen, K. (2005). Students' retention of knowledge and skills in differential equations. *School Science and Mathematics, 105*(5), 227–239. doi:10.1111/j.1949-8594.2005.tb18163.x.

Laborde, C., & Laborde, J.-M. (1992). Problem solving in geometry: From microworlds to intelligent computer environments. In Ponte (ed.), Mathematical Problem Solving and New Information Technology. Berlin: Springer

Laborde, C. (2001). Integration of technology in the design of geometry tasks with Cabri-geometry. *International Journal of Computers for Mathematical Learning, 6*(3), 283–317. doi:10.1023/A:1013309728825.

Laborde, C., & Laborde, J.-M. (1995). What about a learning environment where euclidean concepts are manipulated with a mouse? In diSessa, A. (Ed.), *Computers for Exploratory Learning*. Berlin: Springer.

Lakoff, G., & Nunez, R. (2000). *Where mathematics comes from.* New York, NY: Basic Books.

Lambertus, A. J. (2010). *Characterizing gender diverse graduate mathematics departments as communities of practice.* Doctoral dissertation.

Lave, J., & Wenger, E. (1991). *Situated learning: Legitimate peripheral participation.* Cambridge, UK: Cambridge University Press. doi:10.1017/CBO9780511815355.

Lee, C. P. (2007). Boundary negotiating artifacts: Unbinding the routine of boundary objects and embracing chaos in collaborative work. *Computer Supported Cooperative Work, 16*, 307–338. doi:10.1007/s10606-007-9044-5.

Lesh, R., Post, T., & Behr, M. (1987). Dienes revisited: Multiple embodiments in computer environments. In Wirsup, I., & Streit, R. (Eds.), *Development in School Mathematics Education Around the World* (pp. 647–680). Reston, VA: National Council of Teachers of Mathematics.

Lim, C. P., & Khine, M. S. (2006). Managing teachers' barriers to ICT integration in Singapore schools. *Journal of Technology and Teacher Education*, *14*(1), 97–125.

Linz, S. J. (1997). Nonlinear dynamical models and jerky motion. *American Journal of Physics*, *65*, 523–525. doi:10.1119/1.18594.

Ljasko, Boljarcuk, Gaj, & Golovac. (2007). *Zbirka zadataka iz matematicke analize*. Belgade, Serbia: Nasa Knjiga.

Makar, K., & Rubin, A. (2007). Beyond the bar graph: Teaching informal statistical inference in primary school. In *Proceedings of the Fifth International Research Forum on Statistical Reasoning, Thinking, and Literacy (SRTL-5)*. Warwick, UK: SRTL.

Meletiou-Mavrotheris, M. (2003). Technological tools in the introductory statistics classroom: Effects on student understanding of inferential statistics. *International Journal of Computers for Mathematical Learning*, *8*(3), 265–297. doi:10.1023/B:IJCO.0000021794.08422.65.

Meletiou-Mavrotheris, M., & Lee, C. (2007). Effects of τechnological tools on introductory statistics students' understanding of sampling distributions. In Murtonen, M., Rautopuro, J., & Väisänen, P. (Eds.), *Learning and Teaching of Research Methods at University* (pp. 95–118). Turku, Finland: University of Turku.

Meletiou-Mavrotheris, M., Paparistodemou, E., & Stylianou, D. (2009). Enhancing statistics instruction in elementary schools: Integrating technology in professional development. *The Montana Math Enthusiast*, *6*(1&2), 57–78.

Merkle. (2006). *Matematicka analiza*. Belgarde, Serbia: CET.

Miller, H. R., & Upton, D. S. (2008). Computer manipulatives in an ordinary differential equations course: Development, implementation, and assessment. Retrieved from http://math.mit.edu/mathlets/wp-content/uploads/cet-published.pdf

MIT Mathlets. (n.d.). Retrieved from http://math.mit.edu/mathlets/

MIT OCW Differential Equations Course. (2010). Retrieved from http://ocw.mit.edu/courses/mathematics/18-03-differential-equations-spring-2010/

MIT Open Courseware. (n.d.). Retrieved from http://ocw.mit.edu

Monaghan, J. (2004). Teachers' activities in technology-based mathematics lessons. *International Journal of Computers for Mathematical Learning*, *9*, 327–357. doi:10.1007/s10758-004-3467-6.

Monoyiou & Gagatsis. (2008). The stability of students' approaches in function problem solving: A coordinated and an algebraic approach. In *Proceedings of the Conference of Five Cities: Nicosia, Rhodes, Bologna, Palermo, Locarno "Research in Mathematics Education"*. University of Cyprus.

Nardi, E., & Iannone, P. (2003). Mathematicians on concept image construction: Single 'landscape' vs 'your own tailor-made brain version'. In N. Pateman, B. J. Dougherty, & J. Zillox (Eds.), *Proceedings of the 27th PME International Conference*, (vol. 3, pp. 365-372). PME.

National Research Council. (2000). *How people learn: Brain, mind, experience, and school*. Washington, DC: National Academy Press.

NCTM. (n.d.). Retrieved from http://www.nctm.org/

Newell, A., & Simon, H. (1972). *Human problem solving*. Englewood Cliffs, NJ: Prentice-Hall.

NOAA. (1995). Ensemble forecasting at NCEP. *National Centers for Environmental Prediction, National Oceanic and Atmospheric Administration*. Retrieved from http://www.emc.ncep.noaa.gov/gmb/ens/info/ens_detbak.html

Noss, R., & Hoyles, C. (1996). *Windows on mathematical meanings: Learning cultures and computers*. Dordrecht, The Netherlands: Kluwer Academic Publishers. doi:10.1007/978-94-009-1696-8.

Nusse, H. E., & Yorke, J. (1996). Wada basin boundaries and basin cells. *Physica D. Nonlinear Phenomena*, *90*(3), 242–261. doi:10.1016/0167-2789(95)00249-9.

Ocak, M. (2005). Mathematics teachers' attitudes toward the computers. *The Turkish Online Journal of Educational Technology*, *4*(3).

Olivero, F. (1999). Cabri-Géomètre as a mediator in the process of transition to proofs in open geometric situations. In: W. Maull & J. Sharp (Eds.), *Proceedings of ICTMT 4*. Plymouth, UK: University of Plymouth.

Özmantar, M., Akkoç, H., Bingölbali, E., Demir, S., & Ergene, B. (2010). Pre-service mathematics teachers' use of multiple representations in technology-rich environments. *Eurasia Journal of Mathematics*, *6*(1), 19–36.

Paparistodemou, E., & Meletiou-Mavrotheris, M. (2010). Engaging young children in informal statistical inference. In C. Reading (Ed.), *Data and Context in Statistics Education: Towards an Evidence-Based Society: Proceedings of the Eighth International Conference on Teaching Statistics* (ICOTS8). Ljubljana, Slovenia: International Statistical Institute.

Paparistodemou, E., & Noss, R. (2004). Designing for local and global meanings of randomness. In *Proceedings of the Twentieth Eighth Annual Conference of the International Group for the Psychology of Mathematics Education* (pp. 497-504). Bergen, Norway: PME.

Paparistodemou, E., & Meletiou-Mavrotheris, M. (2008). Developing young students' informal inference skills in data analysis. *Statistics Education Research Journal*, *7*(2), 83–106.

Paparistodemou, E., Noss, R., & Pratt, D. (2008). The interplay between fairness and randomness in a spatial computer game. *International Journal of Computers for Mathematical Learning*, *13*(2), 89–110. doi:10.1007/s10758-008-9132-8.

Pedoe, D. (1976). *Geometry and the visual arts*. New York: Dover.

Pfannkuch, M. (2006). Informal inferential reasoning. In A. Rossman & B. Chance (Eds.), *Working Cooperatively in Statistics Education: Proceedings of the Seventh International Conference of Teaching Statistics (ICOTS-7)*. Salvador, Brazil: ICOTS.

Pfannkuch, M., Budgett, S., Parsonage, R., & Horring, J. (2004). *Comparison of data plots: Building a pedagogical framework*. Paper presented at the International Association for Statistics Education 2004. Lund, Sweden: IASE.

Polking, J. C. (2005). *Dfield,and pplane.* Retrieved from http://math.rice.edu/~dfield/dfpp.html

Pólya, G. (1945). *How to solve it.* Princeton, NJ: Princeton University Press.

Portides, D. (2008). Representation per se vs. representation as: A useful distinction for mathematics education. In *Proceedings of the Conference of Five Cities: Nicosia, Rhodes, Bologna, Palermo, Locarno "Research in Mathematics Education"*. University of Cyprus.

Pratt, D. C. (1998). *The construction of meanings in and for a stochastic domain of abstraction.* (Ph.D. Thesis). University of London, London, UK.

Pratt, N., & Kelly, P. (n.d.). *Mapping mathematical communities: Classrooms, research communities and master classes.* Retrieved July 3, 2011 from http://orgs.man.ac.uk/projects/include/experiment/kelly_pratt.pdf

Pratt, D. (2000). Making sense of the total of two dice. *Journal for Research in Mathematics Education, 31*, 602–625. doi:10.2307/749889.

Pratt, D., Johnston-Wilder, P., Ainley, J., & Mason, J. (2008). Local and global thinking in statistical inference. *Statistics Education Research Journal, 7*(2), 107–129.

Presmeg, N. C. (1997). Generalization using imagery in mathematics. In English, L. D. (Ed.), *Mathematical reasoning: Analogies, metaphors and images* (pp. 299–312). Mahwah, NJ: Erlbaum.

Rasmussen, C., & Kwon, O. (2007). An inquiry oriented approach to undergraduate mathematics. *The Journal of Mathematical Behavior, 26*, 189–194. doi:10.1016/j.jmathb.2007.10.001.

Rasmussen, C., Kwon, O., Allen, K., Marrongelle, K., & Burtch, M. (2006). Capitalizing on advances in mathematics and K-12 mathematics education in undergraduate mathematics: An inquiry-oriented approach to differential equations. *Asia Pacific Education Review, 7*, 85–93. doi:10.1007/BF03036787.

Rasmussen, C., & Marrongelle, K. (2006). Pedagogical content tools: Integrating student reasoning and mathematics in instruction. *Journal for Research in Mathematics Education, 37*, 388–420.

Rasmussen, C., & Ruan, W. (2008). Using theorems-as-tools: A case study of the uniqueness theorem. In Carlson, M., & Rasmussen, C. (Eds.), *Making the Connection: Research and Teaching in Undergraduate Mathematics.* Washington, DC: The Mathematical Association of America.

Rasmussen, C., Zandieh, M., King, K., & Teppo, A. (2005). Advancing mathematical activity: A view of advanced mathematical thinking. *Mathematical Thinking and Learning, 7*, 51–73. doi:10.1207/s15327833mtl0701_4.

Rasmussen, C., Zandieh, M., & Wawro, M. (2009). How do you know which way the arrows go? The emergence and brokering of a classroom mathematics practice. In Roth, W.-M. (Ed.), *Mathematical representations at the interface of the body and culture* (pp. 171–218). Charlotte, NC: Information Age Publishing.

Reichardt, C. S., & Gollob, H. F. (1997). When confidence intervals should be used instead of statistical tests, and vice versa. In Harlow, L. L., Mulaik, S. A., & Steiger, J. H. (Eds.), *What if There Were no Significance Tests?* (pp. 259–284). London: Lawrence Erlbaum Associates.

Rochowicz, J. A. Jr. (1996). The impact of using computers and calculators on calculus instruction: Various perceptions. *Journal of Computers in Mathematics and Science Teaching, 15*(4), 423–435. Retrieved from http://www.editlib.org/p/15206.

Rott, B. (2011). Problem solving processes of fifth graders: An analysis. In *Proceedings of PME 35*. Ankara, Turkey: PME.

Rott, B., & Gawlick, T. (Manuscript submitted for publication). Sind mathematische heurismen trainierbar? *Unterrichtswissenschaft*..

Rubin, A., Bruce, B., & Tenney, Y. (1990). *Learning about sampling: Trouble at the core of statistics*. Paper presented at the Annual Meeting of the American Educational Research Association. New York, NY.

Rubin, A., Hammerman, J., & Konold, C. (2006). Exploring informal inference with interactive visualization software. In A. Rossman & B. Chance (Eds.), *Working Cooperatively in Statistics Education: Proceedings of the Seventh International Conference of Teaching Statistics* (ICOTS-7). Salvador, Brazil: ICOTS.

Sarra, S. (n.d.). *Derivatives*. Retrieved from http://www.scottsarra.org/applets/calculus/Derivatives2.html

Schau, C., & Mattern, N. (1997). Assessing students' understanding of statistical relationships. In Gal, I., & Garfield, J. B. (Eds.), *The Assessment Challenge in Statistics Education*. Burke, VA: IOS Press.

Schoenfeld, A. (1985). *Mathematical problem solving*. Orlando, FL: Academic Press.

Schoenfeld, A. (1992). Learning to think mathematically: Problem solving, metacognition, and sensemaking in mathematics. In Grouws, D. (Ed.), *Handbook for Research on Mathematics Teaching and Learning*. New York: MacMillan.

Shamatha, J. H., Peressini, D., & Meymaris, K. (2004). Technology-supported mathematics activities situated within an effective learning environment theoretical framework. *Contemporary Issues in Technology and Teacher Education, 3*(4). Retrieved July 2, 2011, from http://www.citejournal.org/vol3/iss4/mathematics/article1.cfm

Sheehan & Nillas. (2010). Technology integration in secondary mathematics classrooms: Effect on students' understanding. *Journal of Technology Integration in the Classroom, 2*(3).

SLU. (n.d.). Retrieved from http://www.slu.edu/classes/maymk/SecantTangent/Secant-Tangent.html

Smale, S. (1967). Differentiable dynamical systems. *Bulletin of the American Mathematical Society, 73*(6), 747–817. doi:10.1090/S0002-9904-1967-11798-1.

Solomon, Y. (2007). Not belonging? What makes a functional learner identity in undergraduate mathematics? *Studies in Higher Education, 32*(1), 79–96. doi:10.1080/03075070601099473.

Sprott, J. C. (1994). Some simple chaotic flows. *Physical Review E: Statistical Physics, Plasmas, Fluids, and Related Interdisciplinary Topics, 50*, R647–R650.

Stake, R. (1995). *The art of case study research*. Thousand Oaks, CA: Sage Publications.

Star, S. L., & Griesemer, J. R. (1989). Institutional ecology, translations and boundary objects: Amateurs and professionals in Berkeley's museum of vertebrate zoology, 1907-1939. *Social Studies of Science*, *19*(3), 387–420. doi:10.1177/030631289019003001.

Steinbring. (1990). The use of chance-concept in everyday teaching - Aspects of a socially constituted epistemology of mathematical knowledge. In J. B. Garfield (Ed.), *Research Papers from the Third International Conference on Teaching Statistics*. Dunedin, New Zealand: University of Otago.

Stohl, H., & Tarr, J. E. (2002). Using multi-representational computer tools to make sense of inference. In D. Mewborn (Ed.), *Proceedings of the Twenty-Fourth Annual Meeting of the North American Chapter of the International Group for the Psychology of Mathematics Education*. Athens, GA: PME.

Struik, D. J. (1950). *Lectures on classical differential geometry*. New York: Dover.

Stylianou, D., & Meletiou-Mavrotheris, M. (2005). Graphical representation of data: The effect of the use of a dynamical statistics technological tool. *Scientia Pedagogica Experimentalis*, *42*(2), 273–294.

Tall. (2009). Dynamic mathematics and the blending of knowledge structures in the calculus. *ZDM – The International Journal on Mathematics Education*, *41*(4), 481–492.

Tall. (2010). *A sensible approach to the calculus*. Paper presented at the National and International Meeting on the Teaching of Calculus. Puebla, Mexico.

Terrell, R. (1995, Summer-Fall). Some software for teaching partial differential equations. *Newsletter for the Consortium for Ordinary Differential Equations Experiments*, 13-15. Retrieved from http://www.codee.org/library/newsletters/Summer-Fall%201995/

Thomson, J. F. (1954). Tasks and super-tasks. *Analysis*, *15*(1), 1–13. doi:10.1093/analys/15.1.1.

TI-. *N*spire™ CAS. (2007). *Dynamical mathematical software from Texas Instruments*. Retrieved from www.education.ti.com

TSM Resources. (n.d.). Retrieved from www.tsm-resources.com

Turner, M. (2010). *The riddle of the Buddhist monk*. Retrieved July 8, 2011, from http://markturner.org/blending.html

Turner, M., & Fauconnier, G. (2002). *The way we think: Conceptual blending and the mind's hidden complexities*. New York: Basic Books.

Tutsch, J. H. (2011). *Ode factory*. Retrieved from http://odefactory.com/

UIUC. (n.d.). Retrieved from www.cs.uiuc.edu/homes/rsamdan2/Rajhans.MSThesis.pdf

UMN. (n.d.). Retrieved from http://www.ima.umn.edu/~arnold/graphics.html

UTK. (n.d.). Retrieved from http://archives.math.utk.edu/visual.calculus/3/graphing.3/index.html

Van Hiele, P. M. (1986). *Structure and Insight*. Orlando, FL: Academic Press.

Van Oers, B. (2001). Educational forms of initiation in mathematical culture. *Educational Studies in Mathematics*, *46*, 59–85. doi:10.1023/A:1014031507535.

Vérillon, P., & Rabardel, P. (1995). Cognition and artifacts: A contribution to the study of thought in relation to instrumented activity. *European Journal of Psychology of Education*, *10*(1), 77–101. doi:10.1007/BF03172796.

Watson, J. M. (2008). Exploring beginning inference with novice grade 7 students. *Statistics Education Research Journal*, *7*(2), 59–82.

Wenger, E. (1998). *Communities of practice: Learning, meaning, and identity*. Cambridge, UK: Cambridge University Press.

West, B., Strogatz, S., McDill, J. M., Cantwell, J., & Hohn, H. (2008). *Interactive differential equations*. Reading, MA: Pearson Addison Wesley. Retrieved from http://www.aw-bc.com/ide

Wilensky, U. (1993). *Connected mathematics - Building concrete relationships with mathematical knowledge*. (Ph.D. Thesis). Massachusetts Institute of Technology, Cambridge, MA.

Wittmann, E. (1975). Matrix strategies in heuristics. *International Journal of Mathematical Education in Science and Technology*, *6*(2), 187–198. doi:10.1080/0020739750060207.

Yaacob, M., Yuzita, S., & Wester, S. (2010). Towards the development of an automated learning assistant for vector calculus: Integration over planar regions. *The International Journal for Technology in Mathematics Education*, *17*(2), 81–86.

Yu, C. H., & Behrens, J. T. (1994). Identification of misconceptions in learning statistical power with dynamic graphics as remedial tool. *ASAProStEd*, 242-246.

Zieffler, A., Garfield, J., delMas, R., & Reading, C. (2008). A framework to support research on informal inferential reasoning. *Statistics Education Research Journal*, *7*(2), 40–58.

Zimmermann, W. (1991). Visual thinking in calculus. In Zimmermann, W., & Cunningham, S. (Eds.), *Visualization in teaching and learning mathematics* (pp. 127–137). Washington, DC: Mathematical Association of America.

About the Contributors

Samer Habre is an associate professor of Mathematics at the Lebanese American University in Beirut, Lebanon, where he has been serving also as the Chairperson of the Department of Computer Science and Mathematics since October 2008. He had earned his BS in Mathematics from the American University in Beirut in 1984, his MS and PhD in Mathematics from Syracuse University (New York) in 1986 and 1991, respectively, and has been teaching at the Lebanese American University since 1993. In 1998, he was awarded a Fulbright Scholarship that he spent at Cornell University (New York), and in 2005, he was a visiting scholar at California Polytechnic State University. Samer Habre's research interests vary between the teaching of mathematics at the university level with visualization aids (particularly dynamical software programs) and research on discrete dynamical systems. He has numerous publications in both fields.

* * *

Heidi Burgiel has been a Professor in the Department of Mathematics and Computer Science at Bridgewater State University since September of 2003. Her professional goal is to serve her community by understanding and communicating mathematics. She is particularly interested in uses of educational technology, and her field of research is (approximately) discrete geometry. She has come to love communicating mathematics in writing. She has assisted MIT OpenCourseWare with their OCW Scholar math courses, and co-authored *The Symmetries of Things*, a book about repeating patterns, with co-authors John Conway and Chaim Goodman-Strauss.

Douglas Butler, after graduating in Mathematics and Electrical Sciences at Cambridge University, and a spell with EMI Records, has specialised in teaching secondary Mathematics. He has served as Head of Mathematics at Oundle School (Peterborough UK), and was Chairman of the MEI Schools project, a leading UK curriculum development project, for 6 years. A keen pianist and dinghy sailor, he is also author of *Using the Internet: Mathematics* (revised July 2005), a major contributor to the ICMI technology study in Hanoi (December 2006), and a lead author of *Autograph* (version 3.3 Mar 2012). He maintains a large Website (TSM Resources) with educational resources in many subject areas, including a wide collection of images and data for the teaching and learning of secondary mathematics. In 2000, he founded the innovative iCT Training Centre, based at Oundle School (near Cambridge, UK) and runs the TSM (Technology in Secondary/College Mathematics) teacher training events all over the UK and overseas. TSM has a further objective now to train trainers, and the 2012 workshop created 53 new trainers. He is a frequent speaker at international mathematics teachers' conferences, and has delivered opening keynotes at: T-Cubed Conference in Denver in 2006, AMESA conference in Port Elizabeth in 2008, MAV in Melbourne in 2010, and was the lead tutor and presetor for the South East Asia Mathematics Contest in Singapore in 2012. He successfully co-edited the first two issues of the new online magazine "MTi" for the ATM in England, and from Jan 2010, he has coordinated the NCETM group to look at methods to Train Trainers in mathematics technology in the UK. He has also run a series of conferences on Technology for Teaching Music.

Ljubica Diković finished Faculty of Mathematics, Belgrade University, and received M.Sc. degree from the Faculty of Science, University of Kragujevac. She is currently in process of preparation of Doctor Thesis on Faculty of Mathematics, Kragujevac University. Now, she works as professor in Business Technical College in Uzice, Serbia. Her research interests focus on high school mathematics supported by Web technologies and learning technologies. She has authored/coauthored more than 30 research papers,presented on national and international conferences, and published in national and international journals.

Bjørn Felsager took a master in Mathematics and Physics from Copenhagen University, Denmark, 1983. He was a visiting fellow at the Mathematics Department at the Cornell University, 1990-1991, where he learned how to write programs in Mathematics under the supervision of members of the Cornell Faulty: John Hamal Hubbard, professor, and Beverly Henderson West, lecturer. He was subsequently a programmer for the third edition of the MultiMat-programs written for the Danish Association of Mathematics Teachers in the High School, 1991-93. He has been a National Coordinator and T^3-instructor in Denmark from 1998 (T^3 = Teachers Teaching with Technology, a program initiated by Texas Instruments Incorporated). He is an author of several text books in Mathematics and Physics for the Danish High School as well as the monograph *Geometry, Particles, and Fields* (Graduate Texts in Contemporary Physics), Springer 1998. He has a wife, 3 children, and 6 grandchildren. He is now retired from his position as a High School Teacher but is still active writing textbooks and giving in-service courses.

Thomas Gawlick received his Ph.D. in pure mathematics in 1996 at Bonn. From 1998, he worked as assistant professor for didactics of mathematics in Vechta. The academic year 2001-2 he spent as "teacher-in-research" at the Oberstufenkolleg Bielefeld (founded by Hartmut von Hentig). In 2003, he was appointed lecturer at Landau, where he completed his habilitation in mathematics education in 2004. Since 2007, Thomas Gawlick holds a chair for secondary mathematics education at Leibniz University, Hannover.His research interests include dynamical geometry, problem solving, and van-Hiele-based curriculum development.

John H. Hubbard is a Professor of Mathematics at Cornell University. His research over the last five years has been devoted to dynamics in several complex variables. He has co-authored four foundational papers in the field and is also writing three books on a very advanced level, one giving a treatment of Teichmüller space and its applications in theorems of Bill Thurston, the second on dynamics of one complex variable, and the third on differential geometry.

Karen Keene is an Assistant Professor of mathematics education. Her research interested lie in two primary arenas. First, she conducts research in undergraduate mathematics education, primarily concerning differential equations teaching and learning. She focuses on how students learn to conceptually understand solutions to first order differential equations and systems of differential equations. Additionally she researches the social construction of mathematical meaning in undergraduate classrooms. Her second area of research is studying secondary teacher education. She studies teacher's content knowledge and how that connects to their teaching and curriculum development.

Chad Lieberman is a doctoral candidate in the Department of Aeronautics & Astronautics at MIT. He is a research assistant for Professor Karen Willcox in the Aerospace Computational Design Laboratory. His research interests include model reduction, inverse problems, uncertainty quantification, and control theory. Currently, he is working on goal-oriented inference. He has developed an algorithm for minimizing the computational cost of inference-for-prediction by exploiting the control-theoretic observability gramian in linear problems. His S.M. degree was in Computation for Design and Optimization in the School of Engineering at MIT. His master's thesis is titled "Parameter and State Model Reduction for Bayesian Statistical Inverse Problems." He holds a S.B. in Aerospace Engineering and a S.B. in Mathematics, both from MIT.

Maria Meletiou-Mavrotheris is an Associate Professor at the European University Cyprus, and Director of the Research Laboratory in ICT-Enhanced Education. She has a Ph.D. in Mathematics Education (University of Texas at Austin, 2000), an M.Sc. in Statistics (University of Texas at Austin, 1994), an M.Sc. in Engineering (University of Texas at Austin, 1998), an M.A. in Open and Distance Learning (UK Open University, 2008), a B. A. in Mathematics (University of Texas at Austin, 1993), and Teacher's Diploma in Elementary Education (Pedagogical Academy of Cyprus, 1990). Her research work focuses on the study of issues related to the teaching and learning of statistics, while at the same time equally contributing to several other areas of mathematics, science, and technology education. She has established a respected research record through several publications in scholarly international journals, and has been able attract considerable funding from national and international sources.

Haynes Miller joined the MIT mathematics faculty as professor in 1986. A graduate of Harvard, he received the Ph.D. from Princeton under the direction of John Moore in 1974. Following assistant professorships at Harvard and Northwestern Universities, he joined the faculty of the University of Washington in 1977, and the faculty of Notre Dame as professor in 1984. Professor Miller is an algebraic topologist. In 1992-93, he served as Chair of the Pure Mathematics Committee. Since 2004, Professor Miller has chaired the Undergraduate Mathematics Committee. He has worked on many educational initiatives, such as the Cambridge-MIT Exchange program, for which he received the Cambridge-MIT fellowship in 2003. Professor Miller was selected by MIT to be a MacVicar Faculty Fellow in 2005, for a ten-year period, for "exemplary and sustained contributions to the teaching and education of undergraduates at MIT." He also received the Graduate Student Council Teaching Award of the School of Science in 2006.

Chris Rasmussen is Professor of Mathematics Education in the Department of Mathematics and Statistics at San Diego State University. He received an undergraduate degree in mechanical engineering, a master's degree in mathematics, and his Ph.D. in mathematics education at the University of Maryland. His research focuses on the learning and teaching of undergraduate mathematics, with a focus on courses that serve as a transition from students' current ways of reasoning to more formal and abstract ways of reasoning.

Robert Terrell is a senior lecturer in the Mathematics Department at Cornell University. He has taught engineering mathematics and other subjects there for 25 years, and was a mechanical engineer previously. He has a Ph.D. in Mathematics from the University of Virginia.

Beverly West earned a BA in physics from Swarthmore College (1961), and an MA in applied mathematics from University of Wisconsin (1963). She taught advanced mathematics and physics at Milton College, Wisconsin, before pursuing a 32 year teaching career in mathematics at Cornell University. As an original member of CODEE (Consortium for Ordinary Differential Equations Experiments), she spent 1996-1997 at Harvey Mudd College, teaching and working with CODEE; she still works with the current incarnation of CODEE (Consortium of Ordinary Differential Equations Educators). West has coauthored a Prentice Hall *Encyclopedia of Mathematics* for schools, two volumes of *Differential Equations: A Dynamical Systems Approach* with John H. Hubbard, and another textbook. *Differential Equations and Linear Algebra* with Jerry Farlow, Jean Marie McDill, and James E. Hall. Simultaneously she has been a co-developer of several software packages: *MacMath*, *Analyzer**, *Interactive Differential Equations*, and *ODE Architect*. She has spoken at dozens of colloquia and conferences, in the U.S. and abroad, and has run many workshops for colleagues. She was invited (1992) to give plenary addresses at the Grand Opening of the Fields Institute, and at ICME7 in Montreal; she was guest editor of the College Math Journal Special Issue on Differential Equations (1994). Since the early 1980s she has focused on exploiting the revolution in computer graphics to bring more understanding, visualization, predictive power, and appreciation of mathematics to more students, especially those not majoring in mathematics. She continues to write, edit, and advocate for the computer graphics approach, particularly for differential equations.

Karen Willcox is Professor of Aeronautics and Astronautics in the Aerospace Computational Design Laboratory at the Massachusetts Institute of Technology. She is also Co-Director of the MIT Center for Computational Engineering and Associate Head of the Department of Aeronautics and Astronautics. She holds a Bachelor of Engineering Degree from the University of Auckland, New Zealand, and masters and PhD degrees from MIT. Before joining the faculty at MIT, she worked at Boeing Phantom Works with the Blended-Wing-Body group. Prof. Willcox's research and teaching interests lie in computational simulation and optimization of engineering systems with two major research focuses. The first is model reduction for large-scale systems with applications in unsteady aerodynamics, flow control, uncertainty quantification, inverse problems, decision under uncertainty, and multifidelity design methods. The second is aircraft system design and optimization with particular emphasis on multifidelity modeling.

Index

www.ingramcontent.com/pod-product-compliance
Ingram Content Group UK Ltd.
Pitfield, Milton Keynes, MK11 3LW, UK
UKHW050946290926
15470UKWH00005B/1661

9 781466 640504